Wastewater Treatment - Emerging Insights and Innovative Approaches

Edited by Jeyakumar Rajesh Banu, Kavitha S. and Yukesh Kannah Ravi

Published in London, United Kingdom

Wastewater Treatment - Emerging Insights and Innovative Approaches
http://dx.doi.org/10.5772/intechopen.1008019
Edited by Jeyakumar Rajesh Banu, Kavitha S. and Yukesh Kannah Ravi

Contributors
Ana Balea, Angeles Blanco, Annette Shoba Vincent, Carlos Negro, Derian Manuel Lerma-Mancinas, Diana Barraza-Jiménez, Elena Fuente, Elva Marcela Coria-Quiñones, Emmanuel Rotimi Sadiku, Everardt Andre Burger, Guillermo Nieto, Hakan Çelikten, Hugo Iván Flores-Hidalgo, Idowu David Ibrahim, Jacques Snyman, Jose Luis Sanchez-Salvador, Julius Musyoka Ndambuki, Kobe Samuel Mojapelo, M. Concepción Monte, Manuel Alberto Flores-Hidalgo, Ramya Ramadoss, Sandra Iliana Torres-Herrera, Sayda Dinorah Coria-Quiñones, Sevil Yıldız, Süheyla Tongur, Williams Kehinde Kupolati, Şule Büşra Zambak

First published in London, United Kingdom, 2026 by IntechOpen
IntechOpen is the global imprint of INTECHOPEN LIMITED, registered in England and Wales, registration number: 11086078, 167-169 Great Portland Street, London, W1W 5PF, United Kingdom

For EU product safety concerns: IN TECH d.o.o., Prolaz Marije Krucifikse Kozulić 3, 51000 Rijeka, Croatia, info@intechopen.com or visit our website at intechopen.com.

British Library Cataloguing-in-Publication Data
A catalogue record for this book is available from the British Library

Wastewater Treatment - Emerging Insights and Innovative Approaches
Edited by Jeyakumar Rajesh Banu, Kavitha S. and Yukesh Kannah Ravi
p. cm.
Print ISBN 978-1-83634-002-7
Online ISBN 978-1-83634-001-0
eBook (PDF) ISBN 978-1-83634-003-4

Meet the editors

Dr. Jeyakumar Rajesh Banu is a Professor and the Head of the Department of Biotechnology at the Central University of Tamil Nadu (CUTN), Thiruvarur, India. He has 25 years of teaching experience at national and international universities. He is the author of 250 publications in Scopus-indexed journals. He secured eleven major research grants and three consultancy contracts. He guided 17 Ph.D. students and 140 postgraduate students in engineering and science, with a focus on environmental engineering and biotechnology. Dr. J. Rajesh Banu is listed among the world's top 2% of scientists by Stanford University in the United States. His research focuses on converting waste into energy, recovering value-added products from waste, biological wastewater treatment, nutrient removal, membrane bioreactors, and microbial fuel cells.

Dr. Kavitha S. is an Assistant Professor at the Department of Biotechnology, Karpagam Academy of Higher Education, Coimbatore, Tamil Nadu, India. She completed her Ph.D. in the Department of Civil Engineering at Anna University, Chennai, India. She has authored about 120 articles in SCI and SCIE journals with a cumulative impact factor of 900. Her publications achieved 7200 citations. She has authored one book, edited three books, and written 14 book chapters. She was ranked among the Top 2% of Scientists Worldwide by Stanford University, USA. Her primary research areas include energy generation from waste-activated sludge, micro- and macroalgae biomass biorefinery, lignin valorization, and microbial fuel cells.

Dr. Yukesh Kannah Ravi is an Assistant Professor at the Center for Organic and Nanohybrid Electronics, Silesian University of Technology, Gliwice, Poland. He received a bachelor's degree in Civil Engineering and master's and Ph.D. degrees in Environmental Engineering from Anna University, Chennai, Tamil Nadu, India. He has authored or co-authored 72 peer-reviewed scientific journals and 21 book chapters. He is also a reviewer and member of the editorial board for various international journals. He was ranked among the Top 2% of Scientists Worldwide by Stanford University, USA. His research focuses on converting various types of waste biomass into bioenergy/biofuels, as well as biological wastewater treatment, biomass biorefinery, and the recovery of value-added products.

Contents

Preface

Currently, researchers are focusing on identifying cost-effective and sustainable approaches to wastewater treatment, not only to address water scarcity but also to reduce environmental pollution and to recover valuable resources. From both economic and ecological perspectives, the implementation of advanced and renewable treatment technologies is crucial for achieving cleaner water and promoting resource circularity. This book offers a comprehensive understanding of wastewater treatment, providing current insights into innovative methods and technologies for contaminant removal, as well as resource recovery for sustainable infrastructure. The information gathered in this book aimed to support a comprehensive understanding of wastewater treatment strategies and resource recovery.

Jeyakumar Rajesh Banu
Department of Biotechnology,
Central University of Tamil Nadu,
Thiruvarur, Tamil Nadu, India

Kavitha S.
Department of Biotechnology,
Karpagam Academy of Higher Education,
Coimbatore, Tamil Nadu, India

Yukesh Kannah Ravi
Center for Organic and Nanohybrid Electronics,
Silesian University of Technology,
Gliwice, Poland

Section 1

Wastewater Treatment Processes

Chapter 1

Bioinformatics Revolution in Wastewater Treatment Process Optimization

Annette Shoba Vincent and Ramya Ramadoss

Abstract

The bioinformatics revolution is transforming wastewater treatment optimization through advances in molecular bacteriology and next-generation sequencing (NGS). This chapter presents a comprehensive examination of how the integration of bioinformatics with molecular bacteriology and NGS are reshaping wastewater treatment by enabling data-driven insights into microbial community structure and function. This chapter explores how bioinformatics pipelines, combined with NGS, facilitate high resolution profiling of microbial populations, treatment performance, and system durability. These capabilities are essential for improving the safety and efficiency of wastewater reuse and for monitoring potential health hazards—thereby addressing global water scarcity challenges. A dedicated section highlights the emerging use of bacteriophages as precision biocontrol agents, supported by genomic analysis that uncovers host specificity, lytic mechanisms, and bacteria bioprobe agents. Current insights into how these tools are being applied globally within the wastewater treatment facilities not only for operational efficiency but also to enable surveillance of pathogenic bacteria, contributing to early warning systems for public health threats. By bridging molecular science with computational analytics, this bioinformatics-driven approach shifts wastewater treatment from reactive maintenance to proactive system management, laying the groundwork for circular economy principles in water resource management and ensuring long-term resilience and public health protection.

Keywords: next generation sequencing (NGS), metagenomics, bioinformatics, bacteriophages, wastewater

1. Introduction

The evolution of wastewater treatment reflects humanity's growing understanding of public health, environmental protection, and technological innovation. From early practices in ancient civilizations such as Mesopotamia to the major infrastructural shifts of the Industrial Revolution [1], wastewater management has progressed into a highly regulated and engineered discipline.

Contemporary wastewater treatment systems are typically designed as multi-stage operations to meet rigorous discharge and reuse standards. These stages are broadly classified into preliminary, primary, secondary, and tertiary treatment [2]. The preliminary phase involves physical processes—such as screening and grit removal—to eliminate bulky materials and safeguard subsequent treatment units. Primary treatment uses sedimentation, sometimes augmented with coagulants, to remove suspended solids and reduce organic load. The secondary stage of wastewater treatment relies on biological processes to effectively break down organic substances and decrease nutrient loads. Tertiary treatment delivers advanced polishing, employing filtration and chemical or physical disinfection to remove pathogens and trace contaminants. Innovations such as membrane technologies and carbon adsorption are increasingly used to address emerging pollutants and enhance water quality.

Secondary treatment typically involves the use of activated sludge systems, where microorganisms play a critical role in breaking down pollutants. These microbes are particularly effective in facilitating processes such as nitrification and denitrification, which remove nitrogen and phosphorus from the wastewater [2]. Biological methods are widely regarded as cost-effective, eco-friendly, and efficient. However, they are also highly sensitive to environmental conditions such as temperature, pH, and dissolved oxygen levels, which can affect microbial activity and overall treatment performance [3, 4]. Furthermore, incomplete degradation of certain compounds can lead to the formation of secondary pollutants. Therefore, a comprehensive understanding of both microbial community composition and function is crucial to interpret microbial dynamics. This microbial insight forms the foundation for optimizing treatment efficacy, improving system resilience, and advancing safe water reuse.

A 2020 study highlights the critical limitations of traditional culture-based and phenotypic methods in microbial analysis, citing their labor-intensive protocols, high resource demands, and, most importantly, their inability to detect more than 90% of bacterial species that remain unculturable [5]. In response to these limitations, molecular bacteriology has emerged as a transformative approach, enabling the exploration of microbial communities at the genetic and genomic levels. By employing deoxyribonucleic acid (DNA)-based techniques—most notably 16S ribosomal ribonucleic acid (rRNA) gene sequencing—researchers can now identify bacteria with atypical phenotypes, rare or slow-growing organisms, and even previously uncultivable taxa.

This molecular shift is powered by advances in bioinformatics, which allow for the processing and interpretation of large-scale sequence data generated through next-generation sequencing (NGS). Through computational pipelines, scientists can classify microbial taxa, predict functional capabilities of microbial consortia, and monitor dynamic changes in microbial populations within wastewater systems. These insights are vital not only for evaluating treatment efficacy but also for identifying microbial indicators of antimicrobial resistance (AMR) and potential health risks in water reuse applications.

Importantly, this chapter expands the role of bioinformatics beyond bacterial identification to include bacteriophages—viruses that infect bacteria—as novel biocontrol agents in wastewater treatment. Recent bioinformatic tools have enabled the discovery and characterization of phages with targeted lytic activity against pathogenic and antibiotic-resistant bacteria. In our recent published research findings, metagenomic and comparative genomic analyses were employed to identify novel phage sequences from wastewater samples and characterize their lytic domains and host specificity [6]. These findings demonstrate how bioinformatics can uncover therapeutic and ecological functions of phages previously hidden in complex microbial environments. The integration of such phage-based strategies, including the design of engineered phages and targeted phage cocktails, opens new possibilities for biofilm control, sludge management, and membrane biofouling reduction. Together, the convergence of molecular bacteriology and phage biocontrol—enabled by bioinformatics—signals a paradigm shift toward precision microbiome management in wastewater systems, advancing both environmental sustainability and public health resilience.

2. Molecular bacteriology and role of NGS in wastewater systems

Microbial communities are important players in the biological processes of wastewater treatment. These complex and dynamic communities made up of bacteria, archaea, fungi, protozoa, and viruses carry out the essential biochemical transformations that reduce organic matter, remove nutrients, and neutralize contaminants in municipal and industrial wastewater. The structure and function of these communities determine the overall effectiveness and stability of wastewater treatment systems. A foundational process in wastewater treatment is the microbial degradation of organic pollutants. Heterotrophic bacteria metabolize carbon-rich compounds in influent wastewater, converting them into simpler end products such as carbon dioxide, water, and microbial biomass. These transformations occur through aerobic and anaerobic respiration pathways and are central to systems such as the activated sludge process, where microbial flocs are maintained in aerated tanks to promote efficient breakdown of organic material [7].

Nitrogen removal in wastewater systems relies on a series of microbially-mediated redox transformations relying on chemolithoautotrophic bacteria. Similarly, phosphorous removal is driven by phosphorous accumulating organisms (PAO). Since its original isolation from activated sludge in wastewater treatment plants, several bacterial PAOs have since been recognized, namely, *Candidatus Accumulibacter* (Types I and II), *Candidatus Accumulimonas*, and *Tetrasphaera* [8, 9]. These bacteria actively accumulate phosphorus at high aggregation capacities.

Besides nutrient removal, microbial communities play an important role in controlling and indicating the presence of pathogens. Fecal coliforms, such as *Escherichia coli*, are widely used as indicators of water quality, serving as proxies for the possible presence of more harmful pathogenic organisms [10]. One of the dominant families of concern is *Enterobacteriaceae*, which includes numerous genera associated with human disease. Studies have consistently demonstrated the presence of pathogenic bacteria in wastewater belonging to the *Enterobacteriaceae family*, including *Salmonella typhi, Enteropathogenic Escherichia coli*, and *Yersinia enterocolitica*. A comprehensive review of various literature cataloged bacterial species isolated from the water environment including wastewater and associated them with specific diseases, noting high concentrations of these pathogens in untreated influents (see **Table 1**).

Technologies for detecting microbial pathogens in environmental systems have evolved significantly over the years. Traditional culture-based methods, although widely used, are increasingly recognized as limited in scope. These methods are often time-consuming, resource-intensive, and unable to cultivate a large proportion of bacterial species—particularly those from complex, mixed microbial communities—thus falling short of being a diagnostic gold standard [21].

In contrast, molecular bacteriology, which relies on nucleic acid-based techniques, has revolutionized the study of microbial communities by enabling detection and characterization of both culturable and unculturable bacteria. Among these, NGS has

Type of bacteria	Disease caused	Concentration [colony forming units (CFU)/ml]	Water environments
Vibrio cholerae	Cholera	10^3–10^7	Riverbed sediment, wastewater
Salmonella typhi	Salmonellosis	~ 10^7	Riverbed sediment, wastewater
Enteropathogenic Escherichia coli	Gastroenteritis	10^1–10^7	Groundwater (borehole), wastewater
Campylobacter jejuni	Gastroenteritis	Variable*	Clinical samples from wastewater
Shigella dysenteriae	Dysentery	~ 10^7	Riverbed sediment
Yersinia enterocolitica	Yersiniosis	Variable*	Clinical samples from wastewater

Table 1.
Types of bacteria present in different water environments [11–20].

emerged as a powerful tool for generating high-throughput genome sequence data, enabling researchers to conduct detailed computational analyses of microbial communities across various environments. Common NGS sequencing platforms (**Table 2**) include Roche, Illumina, Ion Torrent, Sequencing by Oligonucelotide ligation and detection (SOLiD), Oxford Nanopore, and single molecule real-time (SMRT) sequencing by Pacific Biosciences [22].

NGS sequencing technology allows for in-depth studies of microbial diversity, taxonomic classification, and phylogenetic relationships using bioinformatics tools. However, while DNA-based methods are powerful, they cannot distinguish between viable and nonviable microorganisms. This distinction is crucial, especially when assessing the potential health risk posed by pathogenic bacteria. Viability refers to the ability of microorganisms to replicate, sustain metabolic activity, and maintain homeostasis [23]. Since DNA can persist long after a cell has been inactivated, its detection may result in overestimation of microbial richness and misinterpretation of public health risk [24]. In contrast, RNA—due to its instability and short half-life in the environment—serves as a more reliable indicator of microbial viability. For example, the half-life of *Escherichia coli* RNA is approximately 5 minutes, while that of *Bacillus subtilis* ranges from 7 to 15 minutes [25, 26]. Thus, RNA-based detection methods are increasingly being explored for monitoring viable pathogenic bacteria in environmental samples.

The studies show these bacteria can thrive and persist in wastewater systems. Bacterial concentrations can vary significantly depending on environmental conditions, source, treatment process, and even geographic location.

Two widely used NGS approaches—16S rRNA gene amplicon sequencing and shotgun metagenomics—offer high-resolution insights into microbial population structures and dynamics [27]. 16S rRNA gene amplicon sequencing targets the ubiquitous 16S rRNA gene (also found in archaea) to identify and classify microorganisms [28]. Polymerase chain reaction (PCR) amplification using primers that bind to conserved regions to then amplify the variable/hypervariable regions within the gene (which have been described as the DNA fingerprint of bacteria). This technique leverages conserved phylogenetic relationships of bacterial/archaeal 16S ribosomal genes for taxonomic classification, typically at the genus level, and occasionally even at

Manufacturer	Model	Template preparation method	Sequencing method	Sequencer run-time	Data output (per run)	Average read length
Roche	454 FLX Titanium	Microbead based emulsion PCR	Pyro-sequencing	10 h	450 MB	450 bp
	454 FLX+			23 h	700 MB	700 bp
	454 GS Junior			10 h	35 MB	450 bp
Illumina	Illumina GAIIx	Flow cell surface supporting Bridge PCR	Reversible terminator sequencing-by-synthesis	6 days	900–1000 GB	2 × 125 bp
	Illumina HiSeq1000			40 h	250–300 GB	2 × 250 bp
	Illumina HiSeq1500			55 h	13.2–15 GB	2 × 300 bp
	Illumina HiSeq2000			29 h	100–120 GB	2 × 150 bp
	Illumina HiSeq 2500			26 h	32–39 GB	2 × 150 bp
	Illumina MiSeq			8 h	15 GB	2 × 300 bp
	Illumina NextSeq550			12–30 h	120 GB	2 × 150 bp
	Illumina HiSeqX			< 3 days	1.6–1.8 Tb	2 × 150 bp
	Nova Seq 6000			13–44 h	80–6000 Gb	2 × 250 bp

Manufacturer	Model	Template preparation method	Sequencing method	Sequencer run-time	Data output (per run)	Average read length
Life Technologies	SOLiD4	Microbead based emulsion PCR	Sequencing-by-ligation	8 days	260 Gb	35 bp or 50 bp (fragment) 50 bp x 25 bp (paired-end) Up to 50 bp x 50 bp (mate-paired)
	SOLiD5500			8 days	320 Gb	75 bp (fragment) 75 bp x 35 bp (paired-end) Up to 60 bp x 60 bp (mate-paired)
	SOLiD5500xl			7 days	300 Gb	
	SOLiD5500W	Flow chip surface PCR		10 days	320 Gb	
	SOLiD5500xlW			10 days	320 Gb	85 bp (fragment) 75 bp x 35 bp (paired-end) Up to 60 bp x 60 bp (mate-paired)
Life Technologies	Ion PGM	Emulsion PCR on microbeads	Semiconductor dependent sequencing-by-synthesis	7.3 h	1.2–2 Gb	400 bp
Life Technologies	Ion Proton	Emulsion PCR on microbeads	Semiconductor dependent sequencing-by-synthesis	2–4 h	Upto10Gb	200 bp
Pacific Biosciences	PacBioRS	No amplification, SMRTbell template	Single-molecule, real-time DNA sequencing-by-synthesis	2–3 h per cell	400 Mb per cell	8000 bp
Oxford Nanopore	Flongle	No amplification	Nanopore sequencing (nanopore flow-cell passage of DNA)	1 min-16 h	2.8 Gb	> 4 Mb
	MinIONMk			1 min-72 h	50 Gb	
	PromethION			1 min-72 h	14 Tb	
	GridIONMk			1 min-72 h	250 Gb	

Table 2.
Current models of NGS sequencing platforms.

the species level [28]. Key considerations for choosing a sequencing platform include genome size, complexity (G + C content), and required coverage depth and accuracy. Amplicon sequencing favors smaller benchtop sequencers like Roche 454 GS Junior, Illumina MiSeq, and Ion PGM with rapid turnaround time and integral data-analysis workflow.

The Ion Torrent 16S rRNA gene amplicon sequencing workflow simultaneously examines 7 out of the 9 hypervariable regions in the bacterial 16S rRNA gene using PCR primers targeting over 80% of the sequences found in the Greengenes database (**Figure 1**). The Ion Reporter™ software's 16S rRNA workflow module classifies individual reads by aligning them in two-steps, first to the manually curated MicroSEQ® ID library where unaligned reads are then aligned to publicly available Greengenes database [29].

Shotgun metagenomics sequencing can characterize a microbial community through construction of a random total DNA (metagenome) followed by untargeted sequencing [28]. This technique enables comprehensive sequencing of both coding and noncoding genomic regions, providing a robust platform for the species and even strain level identification alongside functional profiling of bacteria, archaea, fungi, and viruses [30]. This sequencing approach provides insights beyond taxonomic identification, enabling the characterization of metabolic pathways, functional gene profiles, antibiotic resistance determinants, and other phenotypic traits. This technology is commonly used on platforms such as Illumina and Oxford Nanopore, with Illumina's systems widely used due to their accuracy. For example, in Illumina's soil metagenomics workflow, libraries pooled for sequencing on the Illumina MiSeq are processed through BaseSpace Sequence Hub, Illumina's cloud-based bioinformatics platform.

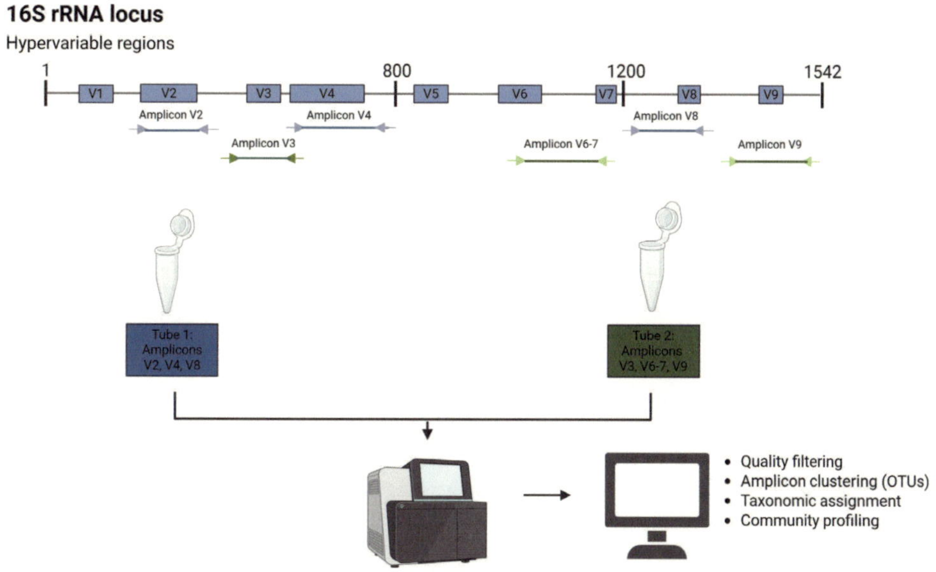

Figure 1.
E. coli *16S rRNA gene with hypervariable regions. The variable regions within* Escherichia coli *16S rRNA gene are indicated in blue squares. Primers (blue arrow lines) targeted V2, V4, and V8 amplicons in a single multiplex PCR tube, while in a separate tube, primers (green arrow lines) targeted amplicons V3, V6–7, and V9. Given the clonal nature of massively parallel NGS, each PCR amplification product and subsequent sequencing read represents a single bacterium within a mixed population. This enables culture-independent characterization of bacterial communities. Simultaneously surveying 7 of the 9 hypervariable regions on the 16S rRNA gene enhances sensitivity and species-level discrimination. [5]. Created in BioRender. Shoba (2025) https://BioRender.com/jl13jce*

In addition to the proprietary data-analysis workflows supported by sequencing platforms, open-source bioinformatics packages and analysis pipelines offer robust alternatives for extracting meaningful insights from sequencing data. CloVR-metagenomics [31], Galaxy webserver for metagenomics [32, 33], IMG/M [34, 35], MetAMOS [36], MG-RAST [37, 38], RAMMCAP [39], and SmashCommunity [31] are some of the widely used metagenomics pipelines. QIIME2 [40], DADA2 [41], Mothur [42], LotuS [43], ASAP 2 [44], ampliseq [45], and EasyAmplicon [46] are the common pipelines used for amplicon sequence data processing.

In the context of water treatment systems, 16S rRNA gene amplicon sequencing and metagenomics data can be computationally analyzed to discern bacterial population in treated water samples [27]. Sequenced data of water samples collected from water-treatment plants, generated from sequencing methodologies, can be put to diversity, taxonomic classification, and phylogenetic studies using QIIME2 tool to get an understanding of the impact of the water-disinfection process and allowed for monitoring of bacterial community richness and diversity at genomic level. Such analyses provide valuable insights into the effectiveness of treatment protocols and the ecological shifts in microbial communities before and after treatment.

3. Microbial community profiling for process monitoring and optimization

Traditional methods for monitoring wastewater treatment primarily rely on physical and chemical parameters like chemical oxygen demand (COD), biochemical oxygen demand (BOD), and total organic carbon (TOC) but provide incomplete understanding of the underlying biological processes. Although microbial communities are the key functional agents in biological wastewater treatment, they are typically not monitored directly in conventional process control. Understanding how microbial communities behave during wastewater treatment is critical for both pathogen control and process optimization. Wastewater treatment systems are designed to significantly reduce microbial pathogens, including fecal indicator bacteria and disease-causing organisms such as *Salmonella*. It is during the primary treatment; significant microbial reduction occurs. It has been reported that the primary treatment stage can remove 25–75% of total suspended solids and up to 50–60% of fecal coliform bacteria, with pathogen removal strongly associated with the sedimentation of solids during this stage [47]. As treatment progresses to biological stages such as activated sludge or suspended growth systems, further microbial elimination is achieved. According to reference [48], 90–99% reductions in fecal coliforms and Salmonellae were documented, attributed in part to low-oxygen environments that inhibit microaerophilic pathogens. When evaluating the efficacy of anaerobic digesters and activated sludge systems in wastewater treatment plants within the municipal and agricultural settings, it was reported that anaerobic digestion removed about 90% of fecal coliforms, while aerobic activated sludge systems achieved even higher removal efficiencies of around 97% in full-scale treatment plants [49]. Additional processes such as membrane bioreactors can further limit bacterial load, creating cleaner effluent streams [48]. However, these studies primarily employed culture-based enumeration and microscopy to assess efficacy—techniques that, as discussed earlier, lack robustness, and may not accurately reflect the microbial community. While microscopy and culture-based assays remain commonly used in wastewater monitoring, they have significant limitations—particularly in specificity, quantitation, and the ability to detect non-culturable organisms [48].

Disinfection represents a critical final barrier in pathogen control. Chlorination, UV radiation, and ozonation are common methods employed during the tertiary treatment stage [50]. Among these, chlorination remains the most widely used due to its cost-effectiveness and established efficacy. Chlorination has been utilized in water treatment for over a century [48]. It is imperative for successful disinfection using chlorination that dosing and contact times are adjusted based on the characteristics of the influent [50]. While this method is highly effective against many bacteria—often reducing fecal coliforms to undetectable levels—it is less successful against certain protozoa such as Giardia and Cryptosporidium [50]. Moreover, although chlorinated bacteria may no longer be culturable, they can remain viable and pose public health risks [51]. Furthermore, a major concern associated with chlorination is the formation of disinfection by-products (DBPs), including trihalomethanes (THMs) and haloacetic acids (HAAs), which result from chlorine reacting with organic matter [52, 53]. These compounds are persistent and can lead to adverse health effects such as congenital defects, stillbirth, and certain cancers [54, 55]. Additionally, antibiotic-resistant bacteria may be reactivated after chlorination, undermining the safety of the treated effluent [56].

UV and ozonation are gaining attention despite their higher operational costs and complexity. Ozonation, while effective, is more likely than chlorination to induce a viable but non-culturable (VBNC) state in organisms such as *Escherichia coli* and *Pseudomonas aeruginosa* [57]. UV radiation, although comparable in cost to chlorination, presents challenges in dosage calibration and has demonstrated limited effectiveness against antibiotic-resistant bacteria. Studies have observed minimal reductions in resistant coliform populations following UV exposure [58]. Studies discussed in Ref. [59] similarly found no significant decline in tetracycline-resistant bacteria post-treatment. While some reduction is possible, UV treatment alone may be insufficient for resistant strains [59].

Despite drawbacks, chlorination remains in widespread use globally due to its affordability and familiarity. However, growing concerns about DBPs and microbial resistance are prompting the exploration of more sustainable alternatives. With sustainability becoming a central focus, alternative disinfection strategies are being explored to complement or replace conventional methods. In arid regions like Qatar, where desalination dominates water supply, enhancing the sustainability of wastewater reuse is essential. Treated wastewater, for instance, has been identified as a potential resource for aquifer recharge, offering a low-impact and cost-effective solution for augmenting groundwater quality [60]. These innovations underscore the potential of integrating microbial community profiling with sustainable disinfection technologies. Profiling helps track shifts in microbial populations in response to treatment processes, thereby informing process control and optimization. As global water demands grow and environmental concerns intensify, embracing such approaches will be essential to building resilient, safe, and sustainable wastewater treatment systems.

NGS and its bioinformatics analysis is crucial for wastewater microbial profiling that provides a high-resolution, comprehensive, and culture-independent solution. 16S rRNA gene amplicon sequencing and shotgun metagenomics (i) allows for the identification and quantification of bacteria, archaea, viruses, and eukaryotes present in wastewater, down to the species or even strain level; (ii) enables pathogen surveillance (e.g., *Escherichia coli*, *Enterococcus*) and monitoring of antimicrobial resistance genes (ARGs); (iii) interpret wastewater treatment efficiency by Identifying prevalence of specific functional indicator microbes that signal efficiency or dysfunction of treatment processes; and (iv) aids in tracking microbial community shifts in response to operational parameters.

A standard NGS bioinformatics analysis pipeline (**Figure 2**) for analyzing either 16S rRNA amplicon sequencing data or metagenomic sequencing data derived from wastewater microbial community profiling include:

i. Quality control (QC) and preprocessing of sequenced reads

In this step, the overall quality of the raw sequencing reads is evaluated. Key metrics are checked to identify potential issues like per-base quality score, per-sequence quality score, sequence length distribution, GC content distribution, N content, adapter content, overrepresented sequences, duplication level etc. that arise during the sequencing process or library preparation. Preprocessing of sequenced reads include read trimming, removal of adapters/barcode, merging (16S rRNA amplicon sequencing) or read trimming, host sequence removal, and filtering for reads belonging to required taxonomy (metagenomic sequencing).

a. Metagenome sequencing: assembly-based analysis or read-based analysis.

- Assembly-based analysis: this process is done to reconstruct the complete or near-complete genomes of individual microorganisms present in a complex environmental sample using short sequencing reads. Firstly, the short quality-controlled raw shotgun metagenomic reads are computationally joined based on overlapping sequences by using assembly algorithms (MEGAHIT, metaSPAdes etc.) to generate longer contiguous sequences called "contigs." Next, contigs are further linked to form "scaffolds" based on available orientation and distance information (paired-end or

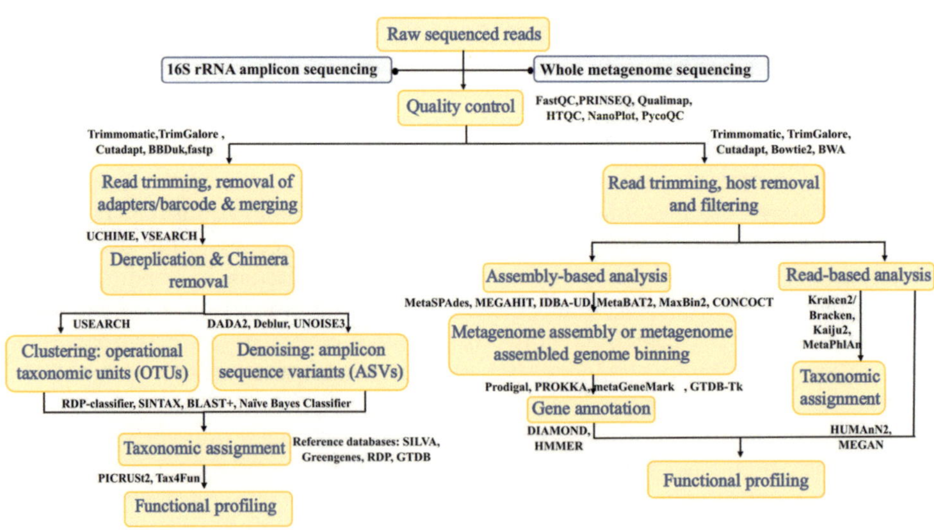

Figure 2.
Standard analysis workflow for amplicon and metagenomic sequencing data. The NGS sequencing of wastewater generates raw sequenced reads data either by whole genome sequencing (WGS) or 16S rRNA amplicon sequencing. The orange boxes indicate each analysis workflow step, with available bioinformatics tool/packages used for the step listed next to arrows.

mate-pair). Further binning algorithms (MetaBAT, MaxBin, CONCOCT) group contigs/scaffolds that likely belong to the same microbial genome based on characteristics like sequence composition, coverage depth and co-abundance patterns. Metagenomic assembly and binning results in a collection of "Metagenome-Assembled Genomes" (MAGs), which are draft genomes of the prevalent microbes in the sample. MAGs are further used for downstream analysis of gene annotation (prediction of genes inherent within the MAGs) followed by functional profiling. This analysis allows for a deeper understanding of the microbial community composition and the functional potential.

- Read-based analysis: instead of proceeding with assembly, the individual reads are directly taxonomically classified using comparison algorithms that compare these reads against reference databases of known microbial genomes (e.g. NCBI RefSeq). This results in an abundance table with the identified taxa and their estimated relative abundances within sample. Apart from taxonomic profiling, the individual reads could also be used directly for functional profiling to understand the metabolic capability of the inherent community by using functional prediction algorithms that align these reads against reference functional databases (KEGG, eggnog, InterPro, etc.) with known genes, enzymes/protein, and metabolic pathways.

b. 16S rRNA amplicon sequencing: clustering of operational taxonomic units (OTUs) or denoising amplicon sequence variants (ASVs)

The QC passed preprocessed amplicon sequences are either processed as OTUs or ASVs that represent two primary strategies for grouping amplicon sequences to differentiate microbial taxa. While they both contribute for characterizing microbial diversity, they differ significantly in their approach and level of resolution.

- OTU clustering is done to categorize highly similar amplicon sequences based on a predefined sequence similarity threshold (mostly 97%) into discrete OTUs with each OTU presumed to represent a distinct microbial group such as a species or genus. For example, two amplicon sequences that are 97% or more identical are grouped into the same OTU.

- In case of ASVs, instead of clustering, initial denoising of amplicons is performed to produce unique and sequencing error corrected high-fidelity ASVs. Even single nucleotide amplicon sequence differences are identifies using ASVs without any similarity threshold, providing a much higher resolution view of genetic diversity that allows for the detection of very closely related species or even strains.

The taxonomic labels (kingdom, phylum, class, order, family, genus, species) of the resulting OTUs/ASVs are assigned by comparing with reference databases like Greengenes, SILVA, RDP, GTDB, and so on using alignment or k-mer based method. Finally, an OTU/ASV abundance table is generated that could be used for further downstream diversity analysis.

ii. Diversity analysis: diversity metrics to understand community structure

a. Alpha (within-sample) diversity: this metric refers to the diversity within a specific sample. It quantifies the richness (number of distinct taxa) and/or evenness (level of even distribution of the taxa) within the sample. Common alpha diversity metrices include Chao1 index, Shannon index, Simpson index, and so on.

b. Beta (between-sample) diversity: these metrics are either similarity or dissimilarity measures of microbial communities to compare different samples that can be qualitative (based on presence or absence) or quantitative (considering relative abundance). The results are visualized using ordination plots like Principal Coordinate Analysis (PCoA) or Nonmetric Multidimensional Scaling (NMDS), where a higher value of dissimilarity measures indicates greater dissimilarity between samples, and a higher value of similarity measures indicates more similarity between samples. The main aim of beta diversity measure is to identify distinct community types, track community shifts, and identify factors that induce variations in community composition. Common beta diversity metrices include Jaccard Distance (qualitative), Bray-Curtis Dissimilarity (quantitative), Unweighted UniFrac (qualitative), and Weighted UniFrac (quantitative).

iii. Statistical analysis to identify differentially abundant taxa or genes: to identify specific microbes or functional capabilities that are significantly different between different conditions (e.g. treated vs. untreated wastewater). The taxonomic abundance table resulting from previous analysis steps along with sample metadata details are input into statistical tools like DESeq2, edgeR or tests like ANOVA, t-tests, Kruskal-Wallis Test, and so on to generate tables, bar plots, heatmaps, or volcano plots to highlight the key taxa or functions that differ significantly between conditions.

iv. Network analysis to pinpoint co-occurrence patterns between microbial taxa or genes.

Instead of examining individual taxa or genes, network analysis enables us to visualize and quantify complex interrelationships and associations between different components within a microbial community (co-occurrence patterns). Typical steps involved in network analysis include

- Calculation of pairwise correlations (e.g., Pearson, Spearman, Kendall) between all pairs of taxa or genes across all samples using tools like SPIEC-EASI, CoNet, and so on.

- Estimation of an adjacency matrix (1-indicate significant link and 0-indicate no link) after filtering out weak or nonsignificant correlations based on a threshold value.

- Network construction based on the adjacency matrix.

- Network visualization using tools like Cytoscape, Gephi, and so on.

- Network property analysis by calculating the network topology metrices like clustering coefficient, degree, modularity, and so forth.

Network analysis aids to infer potential ecological interactions, identify marker species/genes pivotal for maintaining community structure or function, reveal closely interacting microbial consortia, indicate shifts in community stability or response to ecological factors, and generate testable hypotheses about microbial ecosystem functioning.

4. Phage-based biocontrol

Phages belong to the family of viruses that infect bacteria and use them as host to survive. This characteristic of phages makes them a potential biomonitoring tool to detect different bacterial communities in environmental samples such as wastewater. Phage tail fibers, base plates, and endolysins are the primary components of phage infection machinery. Phage-bacterial host interaction occurs through a reversible attachment of the tip of the phage long tail fibers to bacterial cell membrane components like lipopolysaccharides or the porin proteins [61]. When a phage latches onto the bacterium's outer membrane, it transmits a recognition signal to its baseplate. This makes the short tail fibers extend and attach to the outer core of bacterial cell-membrane lipopolysaccharides [62, 63]. Following the attachment of tail fibers, the outer tail sheath contracts, allowing the inner tail tube to penetrate through the bacterial membrane to release viral DNA into the bacterium [64, 65]. The mechanism of phage-host interaction is specific for different bacterial communities, which makes phages a perfect tool for bacterial species elimination in the environmental samples of interest. The nature of virus host interaction is mediated by cell surface receptors, co-receptors, and other molecules depending on the specific host and virus [66, 67].

In a 2015 study, two novel *Escherichia coli* bacteriophages were isolated and identified and further used in wastewater treatment processes [68]. The water from that the bacteria were isolated was obtained from a wastewater treatment plant in Iran, and the bacteriophages were collected from the Zayandehrood River in Iran. According to another study, *Escherichia coli* O157 phages were isolated and identified from wastewater and subsequently used for medical purposes [69]. Another study reported isolation of *Salmonella* bacteriophages from swine lagoons; however, the *Salmonella* strains used for enrichment of the phages were not isolated from the swine lagoon but rather lab strains [70]. In a similar study, different strains of *Salmonella* bacteria strains such as *Salmonella agona*, *Salmonella typhimurium*, and *Salmonella enteritidis* were used for isolation of *Salmonella*-specific phages from wastewater [71].

Due to immense diversity among the phages, it becomes difficult to get a comprehensive understanding of families they belong to by sequencing study. Unlike bacteria, viruses lack the ubiquitous 16S rRNA gene, which makes the identification more complex. Metavironmics (viral metagenomics) is conducted namely using shotgun WGS to profile viral diversity [72]. However, limitations exist due to the low availability of fully sequenced phage genome sequences on public databases alongside the existence of prophages (phage DNA embedded within bacterial genomes) leading to misclassification or inability to be recognized [72]. In recent years, the phage genome sequencing efforts have improved alongside bioinformatics tools specific to phages and prophages such as Phage-Phiser. The recommended methodological flow for viral diversity requires preprocessing, assembly, followed by annotation and profiling. Preprocessing involves using tools like FastQC and PrinSeq to trim adapters and filtration of quality reads from the sequencing data to remove low quality reads. Sequencing reads are put together into viral contigs using assemblers such as Velvet or SPAdes. Alternatively, KAIJU tool aids in taxonomic classification of metagenomic WGS sequencing reads by comparing a reference database of viral and other microbial protein sequences.

Phage proteins, particularly tail fibers and lysins, have potential to be developed into tools for microbial detection and control when integrated into wastewater treatment systems. A bioinformatics-based methodological approach can be employed to identify bacteriophage-derived tail fiber and lysin proteins with potential applications in wastewater surveillance and microbial control. This strategy leverages the host-specific binding properties of tail fiber proteins for bioprobe development, and the bacteriolytic activity of lysins for the discovery of novel biocidal agents. A possible methodology is described below (**Figure 3**).

Step 1—Extraction of target phage protein sequences: Protein sequences associated with bacteriophages infecting members of the family common microbial contaminants in wastewater, for example, *Enterobacteriaceae*—specifically *Salmonella*, *Escherichia*, *Erwinia*, *Shigella*, *Klebsiella*, *Enterobacter*, *Citrobacter*, *Morganella*, *Proteus*, *Yersinia*, and *Serratia*, can be extracted from metagenomic data processed through the KAIJU taxonomic classification pipeline. Sequences can be selected irrespective of read abundance in order to ensure a comprehensive capture of low-abundance but potentially functionally relevant viral taxa.

Step 2—Curation of annotated phage protein sequences: Phages identified using viral protein databases such as NCBI RefSeq Viral Protein Database can then be retained for downstream analysis. Functional annotation should be used to filter for two principal protein types of interest: (1) *Tail-associated proteins*, including those annotated as Tail

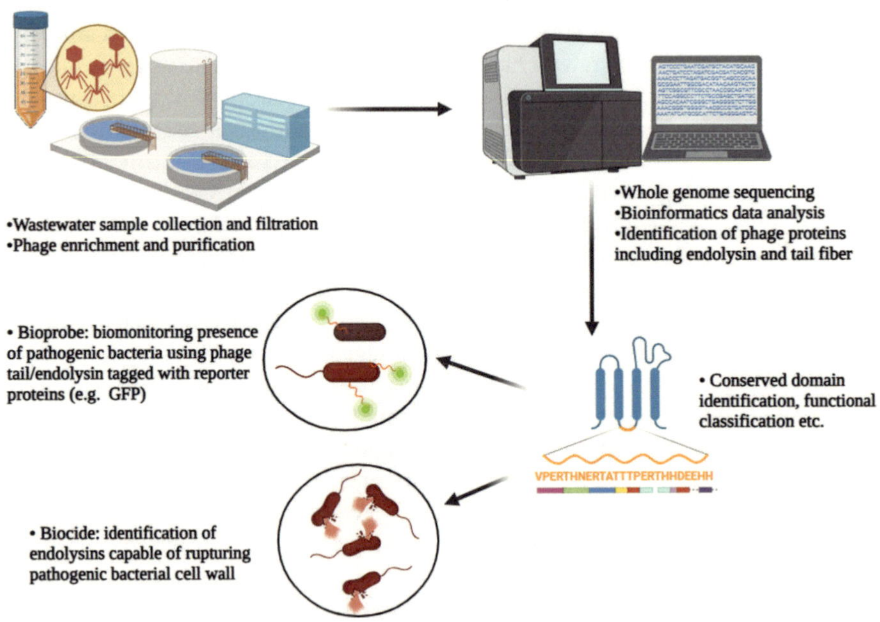

Figure 3.
Schematic workflow describing development of bacteriophage tail fiber or endolysin based biomonitoring and biocidal agents. Created in BioRender. Ramadoss (2025) https://BioRender.com/h9momgr.

Fiber, L-shaped Tail Fiber, Phage Tail Fiber Protein, Tail Spike, Baseplate Wedge Tail Fiber, Central Tail Fiber, Short Tail Fibers, Proximal Tail Fiber, Distal Tail Fiber, Host Specificity Protein, Long Tail Fiber (Proximal and Distal Subunits), and hypothetical or putative tail fibers and (2) *Lysin proteins*, annotated as Lysin, Endolysin, Putative Lysin, or Putative Endolysin. This filtering will result in two curated protein datasets: a tail fiber dataset suitable for bioprobe-related applications and a lysin dataset for biocidal candidate discovery.

Step 3—Functional classification and phylogenetic analysis: To assess the evolutionary relationships and potential functional diversity of the retrieved sequences, phylogenetic trees can be constructed using the Maximum-Likelihood (ML) method. Separate trees will then be generated for the tail fiber and lysin protein datasets. In addition, principal component analysis (PCA) should be performed on the amino acid sequences using the QIIME2 platform [40] to identify clusters of proteins potentially associated with broad host-range activity. Protein clusters spanning related microbial genera being targeted will be prioritized, based on the hypothesis, and these may offer enhanced utility as generalist probes or wide-spectrum lysins.

Step 4—Identification of conserved protein domains: A conserved domain analysis can then be conducted to identify motifs associated with host interaction or lytic activity: For tail fiber proteins, protein sequences containing at least one of the domains of interest will be identified, and the corresponding phage genomes should be retrieved for contextual analysis and potential engineering. For lysin proteins, the most frequently represented domains such as *lyz_endolysin_autolysin* and *endolysin_R21-like* will likely be identified. Proteins carrying these domains were flagged as candidates for further functional validation and antimicrobial assessment.

This methodology provides a structured approach to identifying phage-derived proteins with dual utility: *Bioprobe design*: Tail fiber proteins possessing conserved receptor-binding domains may serve as bio-recognition elements in the construction of phage-based or synthetic biosensors for detecting specific bacterial genera in complex wastewater matrices. *Biocide development*: Lysin proteins exhibiting conserved peptidoglycan-degrading domains may be developed into novel antimicrobials capable of targeting resistant or persistent bacterial strains in wastewater environments. This pipeline is adaptable to other metagenomic contexts and taxonomic targets and can be integrated into broader frameworks for microbial surveillance and bio-intervention strategies in environmental biotechnology.

5. Integrated case studies: NGS and wastewater applications in practice

Application of NGS has been increasingly employed as it is a valuable tool to characterize microbial communities in the wastewater treatment process and optimization [73]. In a 2018 study, distinct differences were observed within anaerobic digesters from biogas plants and sewage treatment plants. Differences in both microbial community composition alongside functional potential was observed. The microbial diversity was found to be greater in sewage treatment plants, while the core microbial community members in the biogas plants exhibiting tighter metabolic integration [74]. While the study attributed the differences to the composition of the influent, these results provide an improvised strategy of treatment that improved efficacy and cost efficiency [74].

The recent COVID-19 pandemic resulted in wastewater surveillance emerging as a powerful, noninvasive tool for monitoring community health. This systematic analysis of wastewater samples was recognized as an opportunity by several countries, including the United States, which subsequently expanded surveillance targets to include indicators such as antimicrobial resistance [75]. Through molecular techniques such as real-time-quantitative PCR (RT-qPCR) and NGS, countries were able to detect pathogens, antimicrobial resistance genes, and other biomarkers at the population level [76]. Countries such as Belgium, Australia, and the United States rapidly deploying wastewater-based epidemiology to track SARS-CoV-2 transmission [76]. In Belgium, detection of SARS-CoV-2 RNA in wastewater provided early warning signals of rising infection rates and helped guide public health interventions. However, upon analysis of several research articles on wastewater surveillance programs, only 2 out of 44 targets mentioned were identified as bacteria, namely, *Escherichia coli* and *Salmonella* species [77]. In a recent review reporting the findings from a workshop held to identify current challenges and future goals for wastewater surveillance, discussions on existing or previous surveillance programs identified 16 targets, with majority related to AMR and fecal associated bacteria [75]. Key challenges and barriers identified include the selection of new surveillance targets, the correlation between human infections and laboratory findings, and issues related to methodology and data normalization [75]. When selecting new targets, there was broad consensus that bacterial indicators should meet three criteria: pose a public health concern, be detectable through the surveillance system, and provide actionable data to inform specific health interventions [75].

An approach of amplicon generation using PCR for detection was described in a 2004 study where PCR-based methods were optimized to allow for rapid detection of *Campylobacter jejuni*, *Campylobacter coli*, and *Yersinia enterocolitica* serovar 0:3 in wastewater samples (site). While the method is effective for known pathogens, unknown pathogens within the sample remain undetected. Unbiased metagenomic approaches allow for comprehensive identification of previously uncharacterized microbial communities within a single wastewater sample. When coupled with genomic sequencing, wastewater monitoring can also facilitate the detection and tracking of emerging pathogens in real time [75].

Wastewater surveillance is being increasingly explored for tracking climate-sensitive pathogens, such as those associated with rising temperatures and flooding events. It has been reported that all water-borne pathogens that have links to climate change are detected in wastewater. Various detection methods, including metagenomics, qPCR, and immunoassays, are being employed to identify pathogens like *Vibrio*, *Leptospira*, and *Cryptosporidium* in wastewater systems. These examples underscore the versatility of wastewater surveillance as a tool not only for disease monitoring but also for assessing environmental and societal trends [76]. However, validation of the methods and normalization of data needs to be improved.

6. Challenges and future directions

Despite its transformative potential, the application of NGS and bioinformatics in wastewater microbiome profiling and pathogen detection is not without its inherent challenges and limitations. The challenges range from the inherent properties of the sample itself to the intricacies of data analysis and wider deployment obstacles. NGS technologies generate massive amounts of data and require robust and specialized bioinformatics pipelines for processing and interpretation. This necessitates significant bioinformatics expertise, which can be a limiting factor for many laboratories. Challenges in *de novo* assembly for novel sequences arise from the incompleteness of microbial databases, which can hinder correct assembly and binning [78]. This often results in a significant percentage of unique reads remaining unclassified taxonomically [79]. Furthermore, linking specific ARGs or pathogenic genes to their precise bacterial taxa, especially for plasmid-borne ARGs, remains a complex bioinformatics challenge with shotgun sequencing [78]. While single-cell sequencing protocols show promise in this regard, they are typically time-consuming and resource-intensive [75]. The detection of a particular gene in a metagenomic dataset does not inherently guarantee its expression by the bacteria, complicating functional interpretations [75]. NGS also has a restricted capacity to detect low-frequency variations, which can be critical in situations involving multiple virus variants coexisting in quasispecies or when there are genetic differences within an organism (heteroplasmy) [80]. Overall, there is a general lack of standardization in bioinformatics methods and workflows across different laboratories, hindering comparability and widespread adoption [80].

Even as NGS becomes more affordable, it remains a significant financial commitment. The expense associated with sequencing equipment and high run costs can still be substantial [73]. Third-Generation Sequencing (TGS) platforms, while offering long reads, typically come with higher costs and higher error rates, necessitating multiple sequencing runs for error correction [78]. These economic barriers and limited competence of scientists in bioinformatics significantly impede the widespread integration of NGS for routine monitoring in wastewater.

Wastewater metagenomics is a rapidly evolving field. New bioinformatics algorithms are enhancing our ability to profile microbial species and predict genes with greater accuracy, leading to better community analysis [78]. In parallel, smart wastewater treatment systems are increasingly incorporating AI-driven optimization for real-time monitoring and control. These AI systems can adapt to changing conditions, significantly improving efficiency and reducing costs. This shift toward AI and automation promises a future of more sustainable, resilient, and economical wastewater management, by automating tasks such as water quality checks and treatment adjustments [81]. Complementing this, cloud solutions (like Google Genomics and Amazon Web Services) provide scalable, accessible ways to host and analyze large next-generation sequencing datasets, often with preconfigured tools, minimizing the need for local hardware. Additionally, specialized platforms from companies like QIAGEN Digital Insights offer comprehensive software for managing bioinformatics pipelines and data at an enterprise scale.

7. Conclusion

This chapter explored emerging innovations in wastewater treatment, emphasizing the power of bioinformatics. This data-driven approach is reshaping how wastewater treatment processes and systems are designed and managed. By leveraging next-generation sequencing, microbial profiling of microbial communities with high resolution can be carried out. This allows researchers to gain critical insights into the efficacy of treatment process, hence improving system durability and allow for more efficient water reuse strategies. Concurrently, application of bacteriophages as targeted biocontrol agents is increasingly gaining traction. Recent progress in phage genomics has enabled the identification of genes responsible for host specificity, lytic mechanisms, and biofilm disruption. This molecular-level understanding allows for the strategic deployment of phages to enhance microbial control and detection in an eco-friendly manner.

Looking ahead, the integration of advanced wastewater treatment processes with sophisticated computational tools paves the way for more sustainable, efficient, and safe water reuse strategies. By harnessing these tools to interpret complex datasets—from metagenomic analyses to AMR surveillance, treatment plants can transition from reactive to proactive management of water quality. The end goal to these efforts is to underscore the change in mindset of wastewater not as waste but as a resource for ecologically responsible urban infrastructure.

Acknowledgements

This work was funded by the National Research Foundation, Singapore (NRF) through Ngee Ann Polytechnic under the RIE2025 plan. We extend our heartfelt gratitude to Dr. Basem Shomar for his invaluable insights and constructive feedback, which greatly enriched the development of this chapter. We are also thankful to Shifa Zhura for her dedicated support and contributions. Their encouragement and expertise were instrumental in shaping this work.

Conflict of interest

The authors declare no conflict of interest.

Author details

Annette Shoba Vincent[1*] and Ramya Ramadoss[2]

1 Ngee Ann Polytechnic, Singapore

2 Mahatma Gandhi Medical Advanced Research Institute (MGMARI), Sri Balaji Vidyapeeth (Deemed to be University), Puducherry, India

*Address all correspondence to: vas5@np.edu.sg

References

[1] Lofrano G, Brown J. Wastewater management through the ages: A history of mankind. The Science of the Total Environment. 2010;**408**(22):5254-5264. DOI: 10.1016/j.scitotenv.2010.07.062

[2] Riffat R, Husnain T. Fundamentals of Wastewater Treatment and Engineering. London: CRC Press; 2022. DOI: 10.1201/9781003134374

[3] Jatoi AS, Memon SA, Memon SQ, Pervez A. Biological wastewater treatment: An overview of microbial community and environmental factors influencing performance. Environmental Science and Pollution Research. 2021;**28**(17):20944-20960. DOI: 10.1007/s11356-020-12222-5

[4] Madakka M, Shettima A, Ibrahim S. Biological wastewater treatment and the factors affecting its efficiency: A review. Journal of Environmental Treatment Techniques. 2019;7(1):10-17. Available from: https://www.jett.dormaj.com/docs/Volume7/Issue1/JETT_Vol_7_Issue_1_Pages%2010-17.pdf

[5] Shomar B, Al-Darwish K, Vincent A. Optimization of wastewater treatment processes using molecular bacteriology. Journal of Water Process Engineering. 2019;**33**:101030. DOI: 10.1016/j.jwpe.2019.101030

[6] Ramadoss R, Al-Shukri M, Shomar B, et al. Substantiation of propitious "Enzybiotic" from two novel bacteriophages isolated from a wastewater treatment plant in Qatar. Scientific Reports. 2022;**12**:9093. DOI: 10.1038/s41598-022-13171-8

[7] Madoni P. Protozoa in wastewater treatment processes: A minireview. Italian Journal of Zoology. 2011;**78**(1):3-11

[8] Fuhs GW, Chen M. Microbiological basis of phosphate removal in the activated sludge process for the treatment of wastewater. Microbial Ecology. 1975;**2**(2):119-138. DOI: 10.1007/bf02010434

[9] Wu L, Ning D, Zhang B, Li Y, Zhang P, Shan X, et al. Global diversity and biogeography of bacterial communities in wastewater treatment plants. Nature Microbiology. 2019;**4**(7):1183-1195. DOI: 10.1038/s41564-019-0426-5

[10] Boutilier L, Jamieson R, Gordon R, Lake C, Hart W. Adsorption, sedimentation, and inactivation of *E. Coli* within wastewater treatment wetlands. Water Research. 2009;**43**(17):4370-4380

[11] Ashbolt NJ. Microbial contamination of drinking water and disease outcomes in developing regions. Toxicology. 2004;**198**(1-3):229-238

[12] Cabral JPS. Water microbiology. Bacterial pathogens and water. International Journal of Environmental Research and Public Health. 2010;7(10):3657-3703

[13] Edberg SC, Rice EW, Karlin RJ, Allen MJ. *Escherichia coli*: The best biological drinking water indicator for public health protection. Journal of Applied Microbiology. 2000;**88**(S1):106S-116S

[14] Fong TT, Lipp EK. Enteric viruses of humans and animals in aquatic environments: Health risks, detection, and potential water quality assessment tools. Microbiology and Molecular Biology Reviews. 2005;**69**(2):357-371

[15] Huq A, Colwell RR, Rahman R, Ali A, Chowdhury MAR, Parveen S, et al. Detection of vibrio cholerae O1 in the aquatic environment by fluorescent-monoclonal antibody and culture methods. Applied and Environmental Microbiology. 1990;**56**(8):2370-2373

[16] Lemarchand K, Lebaron P. Occurrence of *Salmonella* spp. and *Cryptosporidium* spp. in a French coastal watershed: Relationship with fecal indicators. FEMS Microbiology Letters. 2003;**218**(1):203-209

[17] Payment P, Locas A. Pathogens in water: Value and limits of correlation with microbial indicators. Ground Water. 2011;**49**(1):4-11

[18] Rompré A, Servais P, Baudart J, de Roubin MR, Laurent P. Detection and enumeration of coliforms in drinking water: Current methods and emerging approaches. Journal of Microbiological Methods. 2002;**49**(1):31-54

[19] Savichtcheva O, Okabe S. Alternative indicators of fecal pollution: Relations with pathogens and conventional indicators, current methodologies for direct pathogen monitoring and future application perspectives. Water Research. 2006;**40**(13):2463-2476

[20] WHO (World Health Organization). Guidelines for Drinking-Water Quality. 4th ed. World Health Organization; 2011

[21] Barcina I, Arana I. The fate of bacterial pathogens in riverine environments. International Microbiology. 2009;**12**(1):21-28. DOI: 10.2436/20.1501.01.82

[22] Navgire GS, Goel N, Sawhney G, Sharma M, Kaushik P, Mohanta YK, et al. Analysis and interpretation of metagenomics data: An approach. Biological Procedures Online. 2022;**24**(1). DOI: 10.1186/s12575-022-00179-7

[23] Cangelosi GA, Meschke JS. Dead or alive: Molecular assessment of microbial viability. Applied and Environmental Microbiology. 2014;**80**(19):5884-5891. DOI: 10.1128/AEM.01763-14

[24] Li B, Yang Y, Ma L, Ju F, Guo F, Tiedje JM, et al. Metagenomic and network analysis reveal wide distribution and co-occurrence of environmental antibiotic resistance genes. The ISME Journal. 2017;**9**:2490-2502. DOI: 10.1038/ismej.2015.59

[25] Hambraeus G, von Wachenfeldt C, Hederstedt L. Genome-wide survey of mRNA half-lives in Bacillus subtilis identifies extremely stable mRNAs. Molecular Genetics and Genomics. 2003;**269**(5):706-714. DOI: 10.1007/s00438-003-0885-5

[26] Selinger DW, Saxena RM, Cheung KJ, Church GM, Rosenow C. Global RNA half-life analysis in Escherichia coli reveals positional patterns of transcript degradation. Genome Research. 2003;**13**(2):216-223. DOI: 10.1101/gr.912603

[27] Mira Miralles M, Maestre-Carballa L, Lluesma-Gomez M, Martinez-Garcia M. High-throughput 16S rRNA sequencing to assess potentially active bacteria and foodborne pathogens: A case example in ready-to-eat food. Foods (Basel, Switzerland). 2019;**8**(10):480. DOI: 10.3390/foods8100480

[28] Usyk M, Peters BA, Karthikeyan S, McDonald D, Sollecito CC, Vazquez-Baeza Y, et al. Comprehensive evaluation of shotgun metagenomics, amplicon sequencing, and harmonization of these platforms for epidemiological studies. Cell Reports Methods. 2023;**3**(1):100391. DOI: 10.1016/j.crmeth.2022.100391

[29] Thermo Fisher Scientific. Ion reporter™ software: 16S metagenomics workflow user guide. Thermo Fisher Scientific. 2020;**MAN0010799**:1-34. Available from: https://assets.thermofisher.com/TFS-Assets/LSG/manuals/MAN0017772_IonReporter_16SMetagenomics_Workflow_UG.pdf

[30] Quince C, Walker AW, Simpson JT, Loman NJ, Segata N. Shotgun metagenomics, from sampling to analysis. Nature Biotechnology. 2017;**35**(9):833-844. DOI: 10.1038/nbt.3935

[31] Harrington ED, Arumugam M, Raes J, Bork P, Relman DA. SmashCell: A software framework for the analysis of single-cell amplified genome sequences. Bioinformatics. 2010;**26**(23):2979-2980. DOI: 10.1093/bioinformatics/btq564

[32] Giardine B, Riemer C, Hardison RC, Burhans R, Elnitski L, Shah P, et al. Galaxy: A platform for interactive large-scale genome analysis. Genome Research. 2005;**15**(10):1451-1455. DOI: 10.1101/gr.4086505

[33] Pond SK, Wadhawan S, Chiaromonte F, Ananda G, Chung W, Taylor J, et al. Windshield splatter analysis with the galaxy metagenomic pipeline. Genome Research. 2009;**19**(11):2144-2153. DOI: 10.1101/gr.094508.109

[34] Markowitz VM, Chen IA, Chu K, Szeto E, Palaniappan K, Pillay M, et al. IMG/M 4 version of the integrated metagenome comparative analysis system. Nucleic Acids Research. 2013;**42**(D1):D568-D573. DOI: 10.1093/nar/gkt919

[35] Chen IMA et al. IMG/M: Integrated genome and metagenome comparative data analysis system. Nucleic Acids Research. 2017;**45**(D1):D507-D516. DOI: 10.1093/nar/gkt919

[36] Treangen TJ et al. MetAMOS: A modular and open source metagenomic assembly and analysis pipeline. IEEE International Conference on Bioinformatics and Biomedicine Workshops. 2009;**2009**:102-103. Available from: https://ieeexplore.ieee.org/abstract/document/4770094

[37] Meyer F et al. The metagenomics RAST server – A public resource for the automatic phylogenetic and functional analysis of metagenomes. BMC Bioinformatics. 2008;**9**:386. DOI: 10.1186/1471-2105-9-386

[38] Wilke A et al. The M5nr: A novel non-redundant database containing protein sequences and annotations from multiple sources and associated tools. BMC Public Health. 2010;**10**(Suppl 1):S3. DOI: 10.1186/1471-2458-10-78

[39] Li W et al. RAMMCAP: A rapid metagenomic sequence analysis platform. BMC Bioinformatics. 2009;**10**:359. DOI: 10.1186/1471-2105-10-359

[40] Bolyen E et al. Reproducible, interactive, scalable and extensible microbiome data science using QIIME 2. Nature Biotechnology. 2019;**37**:852-857. DOI: 10.1038/s41587-019-0209-9

[41] Callahan BJ et al. DADA2: High-resolution sample inference from Illumina amplicon data. Nature Methods. 2016;**13**:581-583. DOI: 10.1038/nmeth.3869

[42] Schloss PD et al. Introducing mothur: Open-source, platform-independent, community-supported software for describing and comparing microbial communities. Applied

and Environmental Microbiology. 2009;**75**(23):7537-7541. DOI: 10.1128/AEM.01541-09

[43] Hildebrand F, Tadeo R, Voigt AY, Bork P, Raes J. LotuS: An efficient and user-friendly OTU processing pipeline. Microbiome. 2014;**2**(1). DOI: 10.1186/2049-2618-2-30

[44] Tian R, Imanian B. ASAP 2: A pipeline and web server to analyze marker gene amplicon sequencing data automatically and consistently. BMC Bioinformatics. 2022;**23**(1). DOI: 10.1186/s12859-021-04555-0

[45] Straub D, Blackwell N, Langarica-Fuentes A, Peltzer A, Nahnsen S, Kleindienst S. Interpretations of environmental microbial community studies are biased by the selected 16S RRNA (gene) amplicon sequencing pipeline. Frontiers in Microbiology. 2020;**11**. DOI: 10.3389/fmicb.2020.550420

[46] Liu Y, Chen L, Ma T, Li X, Zheng M, Zhou X, et al. EasyAmplicon: An easy-to-use, open-source, reproducible, and community-based pipeline for amplicon data analysis in microbiome research. iMeta. 2023;**2**(1):1-13. DOI: 10.1002/imt2.83

[47] Burton F, Tchobanoglous G, Tsuchihashi R, Stensel HD, Inc M. Wastewater Engineering: Treatment and Resource Recovery. USA: McGraw-Hill Education; 2013

[48] Mara D, Horan NJ. Handbook of Water and Wastewater Microbiology. Netherlands: Elsevier; 2003

[49] Morozzi G, Sportolari R, Caldini G, Cenci G, Morosi A. The effect of anaerobic and aerobic wastewater treatment on faecal coliforms and antibiotic-resistant faecal coliforms. Zentralbl Bakteriol Mikrobiol Hyg B Umwelthyg Krankenhaushyg Arbeitshyg Prav Med. 1988;**185**(4-5):340-349

[50] Ben Ayed L, Sabbahi S. Entamoeba histolytica. In: Rose JB, Jiménez-Cisneros B, editors. Water and Sanitation for the 21st Century: Health and Microbiological Aspects of Excreta and Wastewater Management (Global Water Pathogen Project). (R. Fayer and W. Jakubowski (Eds), Part 3: Specific Excreted Pathogens: Environmental and Epidemiology Aspects—Section 3: Protists). MI, UNESCO: Michigan State University, E.Lansing; 2017. DOI: 10.14321/waterpathogens.34

[51] Chen S, Li X, Wang Y, Zeng J, Ye C, Li X, et al. Induction of Escherichia coli into a VBNC state through chlorination/chloramination and differences in characteristics of the bacterium between states. Water Research. 2018;**142**:279-288. DOI: 10.1016/j.watres.2018.05.055

[52] Li X, Mitch WA. Drinking water disinfection byproducts (DBPs) and human health effects: Multidisciplinary challenges and opportunities. Environmental Science & Technology. 2017;**52**(4):1681-1689. DOI: 10.1021/acs.est.7b05440

[53] Yang X, Shang C, Huang J. DBP formation in breakpoint chlorination of wastewater. Water Research. 2005;**39**(19):4755-4767. DOI: 10.1016/j.watres.2005.08.033

[54] Nieuwenhuijsen MJ, Toledano MB, Eaton NE, Fawell J, Elliott P. Chlorination disinfection byproducts in water and their association with adverse reproductive outcomes: A review. Occupational and Environmental Medicine. 2000;**57**(2):73-85. DOI: 10.1136/oem.57.2.73

[55] Morris RD, Audet AM, Angelillo IF, Chalmers TC, Mosteller F. Chlorination, chlorination by-products, and cancer: A meta-analysis. American Journal of Public Health. 1992;**82**(7):955-963. DOI: 10.2105/ajph.82.7.955

[56] Huang J, Hu H, Tang F, Li Y, Lu S, Lu Y. Inactivation and reactivation of antibiotic-resistant bacteria by chlorination in secondary effluents of a municipal wastewater treatment plant. Water Research. 2011;**45**(9):2775-2781. DOI: 10.1016/j.watres.2011.02.026

[57] Zhang S, Guo L, Yang K, Zhang Y, Ye C, Chen S, et al. Induction of Escherichia coli into a VBNC state by continuous-flow UVC and subsequent changes in metabolic activity at the single-cell level. Frontiers in Microbiology. 2018;**9**. DOI: 10.3389/fmicb.2018.02243

[58] Chang JC, Ossoff SF, Lobe DC, Dorfman MH, Dumais CM, Qualls RG, et al. UV inactivation of pathogenic and indicator microorganisms. Applied and Environmental Microbiology. 1985;**49**(6):1361-1365. DOI: 10.1128/aem.49.6.1361-1365.1985

[59] Umar M. From conventional disinfection to antibiotic resistance control—Status of the use of chlorine and UV irradiation during wastewater treatment. International Journal of Environmental Research and Public Health. 2022;**19**(3):1636. DOI: 10.3390/ijerph19031636

[60] Shomar B, Darwish M, Rowell C. What does integrated water resources management from local to global perspective mean? Qatar as a case study, the very rich country with No water. Water Resources Management. 2014;**28**(10):2781-2791. DOI: 10.1007/s11269-014-0636-9

[61] Yu F, Mizushima S. Roles of lipopolysaccharide and outer membrane protein OmpC of Escherichia coli K-12 in the receptor function for bacteriophage T4. Journal of Bacteriology. 1982;**151**(2):718-722. DOI: 10.1128/jb.151.2.718-722.1982

[62] Aksyuk AA, Leiman PG, Shneider MM, Mesyanzhinov VV, Rossmann MG. The structure of gene product 6 of bacteriophage T4, the hinge-pin of the baseplate. Structure. 2009;**17**(6):800-808. DOI: 10.1016/j.str.2009.04.005

[63] Crowther R, Lenk EV, Kikuchi Y, King J. Molecular reorganization in the hexagon to star transition of the baseplate of bacteriophage T4. Journal of Molecular Biology. 1977;**116**(3):489-523. DOI: 10.1016/0022-2836(77)90081-x

[64] Aksyuk AA, Leiman PG, Kurochkina LP, Shneider MM, Kostyuchenko VA, Mesyanzhinov VV, et al. The tail sheath structure of bacteriophage T4: A molecular machine for infecting bacteria. The EMBO Journal. 2009;**28**(7):821-829. DOI: 10.1038/emboj.2009.36

[65] Kostyuchenko VA, Chipman PR, Leiman PG, Arisaka F, Mesyanzhinov VV, Rossmann MG. The tail structure of bacteriophage T4 and its mechanism of contraction. Nature Structural & Molecular Biology. 2005;**12**(9):810-813. DOI: 10.1038/nsmb975

[66] Dimitrov DS. Virus entry: Molecular mechanisms and biomedical applications. Nature Reviews Microbiology. 2004;**2**(2):109-122. DOI: 10.1038/nrmicro817

[67] Skehel JJ, Wiley DC. Receptor binding and membrane fusion in virus entry: The influenza hemagglutinin. Annual Review of Biochemistry.

2000;**69**(1):531-569. DOI: 10.1146/annurev.biochem.69.1.531

[68] Maal KB, Delfan AS, Salmanizadeh S. Isolation and identification of two novel Escherichia coli bacteriophages and their application in wastewater treatment and coliform's phage therapy. Jundishapur. Journal of Microbiology. 2015;**8**(3):e14945. DOI: 10.5812/jjm.14945

[69] Adibi M, Mobasher N, Ghasemi Y, Mohkam M, Mobasher MA. Isolation, Purification and Identification of *E. coli* O157 Phage for Medical Purposes. DOAJ. DOAJ: Directory of Open Access Journals; 2017. DOI: 10.1111/tips.v3i1.117

[70] McLaughlin MR, Balaa MF, Sims J, King R. Isolation of salmonella bacteriophages from swine effluent lagoons. Journal of Environmental Quality. 2006;**35**(2):522-528. DOI: 10.2134/jeq2005.0080

[71] Akhtar M, Viazis S, Diez-Gonzalez F. Isolation, identification and characterization of lytic, wide host range bacteriophages from waste effluents against salmonella enterica serovars. Food Control. 2013;**38**:67-74. DOI: 10.1016/j.foodcont.2013.09.064

[72] Bruder K, Maiki K, Cooper A, Sible E, Shapiro JW, Watkins SC, et al. Freshwater metaviromics and bacteriophages: A current assessment of the state of the art in relation to bioinformatic challenges. Evolutionary Bioinformatics. 2016;**12s1**:25-33. DOI: 10.4137/ebo.s38549

[73] Chan AW, Naphtali J, Schellhorn HE. High-throughput DNA sequencing technologies for water and wastewater analysis. Science Progress. 2019;**102**(4):351-376. DOI: 10.1177/0036850419881855

[74] Buettner C, Noll M. Differences in microbial key players in anaerobic degradation between biogas and sewage treatment plants. International Biodeterioration & Biodegradation. 2018;**133**:124-132. DOI: 10.1016/j.ibiod.2018.06.012

[75] Philo SE, De León KB, Noble RT, Zhou NA, Alghafri R, Bar-Or I, et al. Wastewater surveillance for bacterial targets: Current challenges and future goals. Applied and Environmental Microbiology. 2023;**90**(1):e0142823. DOI: 10.1128/aem.01428-23

[76] Van Van Poelvoorde L, Karlsson EA, Dupont-Rouzeyrol M, Roosens NHCJ. Can wastewater surveillance enhance genomic tracking of climate-driven pathogens? Microorganisms. 2025;**13**(2):294. DOI: 10.3390/microorganisms13020294

[77] Kilaru P, Hill D, Anderson K, Collins MB, Green H, Kmush BL, et al. Wastewater surveillance for infectious disease: A systematic review. American Journal of Epidemiology. 2022;**192**(2):305-322. DOI: 10.1093/aje/kwac175

[78] Zhang L, Chen F, Zeng Z, Xu M, Sun F, Yang L, et al. Advances in metagenomics and its application in environmental microorganisms. Frontiers in Microbiology. 2021;**12**. DOI: 10.3389/fmicb.2021.766364

[79] Jiang M, Wang AL, Be NA, Mulakken N, Nelson KL, Kantor RS. Evaluation of the impact of concentration and extraction methods on the targeted sequencing of human viruses from wastewater. Environmental Science & Technology. 2024;**58**(19):8239-8250. DOI: 10.1021/acs.est.4c00580

[80] Nafea AM, Wang Y, Wang D, Salama AM, Aziz MA, Xu S, et al. Application of next-generation

sequencing to identify different pathogens. Frontiers in Microbiology. 2024;**14**. DOI: 10.3389/fmicb.2023.1329330

[81] Oladele OK, Federal University of Agriculture. Advanced Wastewater Treatment Systems: Emerging Technologies and Innovations. Germany: ResearchGate; 2024. Available from: https://www.researchgate.net/publication/385214693

Section 2

Applications of Membranes in Wastewater Treatment Plants

Chapter 2

Membrane Technologies in Wastewater Treatment: Current Status, Applications, and Future Perspectives

Hakan Çelikten

Abstract

The growing depletion of water resources, combined with increasing environmental pollution, has highlighted the urgent demand for more effective and sustainable approaches to wastewater treatment. Within this framework, membrane-based technologies have gained significant attention owing to their superior purification performance, compact configurations, and ability to achieve advanced levels of treatment. Functioning as selective physical barriers, membranes are capable of eliminating suspended particles, microorganisms, as well as dissolved organic and inorganic substances from wastewater streams. Depending on the quality of the influent and the targeted treatment objectives, a range of membrane processes, including microfiltration, ultrafiltration, nanofiltration, and reverse osmosis, can be employed. Furthermore, these systems are often incorporated into hybrid configurations and integrated treatment schemes to further improve efficiency and support sustainability. In contemporary membrane processes, dynamic membrane and gravity-driven membrane technologies have gained significant prominence. Furthermore, the integration of artificial neural networks and machine learning applications into membrane process modeling and optimization enables the development of low-energy and sustainable wastewater treatment solutions. This chapter presents a detailed examination of the principles underlying membrane technologies, their practical applications, associated advantages and limitations, as well as insights into their present development and future prospects in the field of wastewater management.

Keywords: membrane technologies, wastewater treatment, hybrid processes, dynamic membrane, gravity-driven membrane, machine learning

1. Introduction

Water covers roughly 71% of the Earth's surface [1], yet the vast majority approximately 97% is saline and confined to the oceans [2]. The remaining freshwater accounts for only 3%. Only 0.5% of freshwater is liquid, found in lakes, rivers, and groundwater; most (2.5%) is frozen in glaciers and permanent snow [2]. UNESCO's (United Nations Educational, Scientific and Cultural Organization) World Water Development Report (2021) indicates that global freshwater demand has increased sixfold over the course of the last century [3]. Today, the significant depletion of freshwater resources and the rising costs associated with them have limited the production capacity of new food products and hindered industrial growth in some countries. This negative impact caused by water scarcity is likely to intensify over the next few years and adversely affect industry worldwide [4].

Urbanization and rising water use in emerging economies cause environmental damage, water scarcity, and viral diseases [5]. Rising water pollution is a major global issue [6] and contributes to many health problems [7]. Poor wastewater treatment and sewage systems discharge pollutants into water sources, endangering ecosystems and human health [7, 8]. Water pollution harms both women's and men's health, causing hormonal changes, reproductive issues, neurological disorders, cancer, and cardiovascular diseases. Chemical, physical, and biological pollutants also lead to premature aging, mental disorders, and other illnesses [9].

Wastewater sources are divided into two main categories: residential (domestic) and non-residential (industrial/commercial) sources (**Figure 1**). Wastewater rich in organic and inorganic wastes consumes the oxygen in freshwater, and its direct discharge causes the death of aquatic life [2]. Domestic wastewater generally contains suspended solids, degradable organic and inorganic compounds, proteins, nutrients (including nitrogen and phosphorus), metals, and disease-causing microorganisms. Non-domestic wastewater, on the other hand, is predominantly generated by industrial processes, agricultural practices, and commercial sectors [1].

Wastewater is water originating from domestic, industrial, and commercial uses; when untreated, it serves no useful purpose and may contain various pollutants [10]. Wastewater consists mainly of water, with the remainder being dissolved or suspended solids of organic, inorganic, and biological origin [11]. Thermal effluents and radioactive substances from energy and nuclear power plants, mining, and similar industrial activities can also be mixed into wastewater [12]. Preventing the negative effects of wastewater and improving water quality are key components of sustainable development and healthy living. Therefore, the regular treatment and disposal of wastewater are of great importance. Wastewater treatment is usually

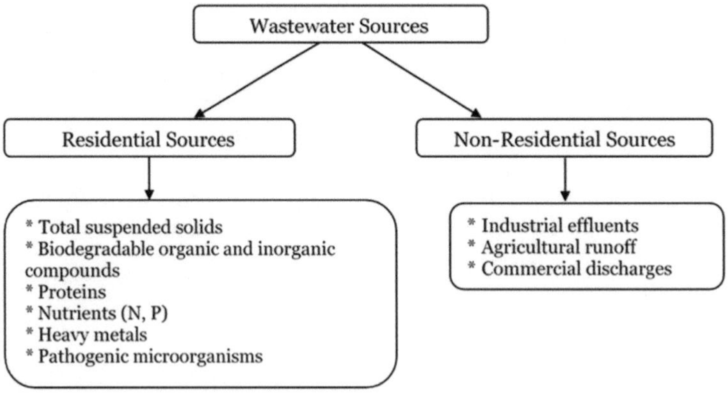

Figure 1.
Classification of wastewater source.

carried out using physical, chemical, and biological methods. For this purpose, aerobic and anaerobic processes [13, 14], membrane technologies [15–17], and advanced hybrid methods [18] are widely applied to different types of wastewater. Treatment techniques differ in their pros and cons, involving not just financial cost but also performance, feasibility, and their impact on the environment. Due to the complex nature of industrial wastewater today, a combination of different methods is often preferred to achieve the desired water quality in the most economical way [19].

Traditional wastewater treatment approaches (e.g., coagulation/flocculation, sedimentation, biological degradation, filtration, adsorption using activated carbon, etc.) require large treatment areas and long processing times. They also generate excessive sludge that entails additional disposal costs and necessitates chemical dosing. Moreover, they often demand supplementary processes to remove agglomerated solids, pose the risk of altering water chemistry due to the use of flocculants, and depend on pumping systems. The performance of low-cost or natural adsorbents is generally limited, while their application at high concentrations is not feasible. In addition, these methods are associated with relatively high capital expenditures and low removal efficiency-to-cost ratios. They also often require post-treatment processes [19, 20]. Traditional treatment techniques are often insufficient for the complete disposal of harmful substances like micropollutants, heavy metals, and persistent organic substances. In contrast, membrane technologies stand out for their compact design, high treatment efficiency, and capacity to remove a wide range of pollutants.

Membrane technology uses semi-permeable membranes to separate fluid components and is applied in water treatment, renewable energy, healthcare, and industries like desalination and gas-liquid separation [21]. Membranes remove contaminants from wastewater with simple operation, easy scalability, and high efficiency while reducing chemical use. Their low energy demand, cost advantage, material recovery, lack of need for phase change, and high water quality enhance sustainability [22].

This study comprehensively addresses the fundamental principles, operating mechanisms, advantages, disadvantages, and applications of membrane technologies used in the treatment of domestic and industrial wastewater in sustainable hybrid systems. The present study aims to examine the factors affecting the efficiency of membrane technology, the functional characteristics of different membrane types, and their current limitations. It also aims to contribute to environmental goals related to improving water quality, recovery, and the efficient use of resources.

2. Membrane technologies and applications

2.1 Membrane types used in wastewater treatment

Urban and industrial wastewater has high sediment content [23] and is also rich in nutrients such as nitrogen and phosphorus. These nutrients promote excessive aquatic plant growth in water bodies and cause eutrophication [24]. When treatment is inadequate, wastewater often accumulates in channels or open drainage systems, where organic matter decomposes, causing foul-smelling gases to form [25].

Therefore, effective and proper treatment of wastewater is extremely important. Various treatment methods are applied to eliminate pollutants from wastewater, including biological processes, membrane processes, chemical precipitation, solvent extraction, oxidation, and adsorption. Today, traditional biological processes used in municipal wastewater treatment face major challenges. These include high energy consumption, excessive sludge production, greenhouse gas emissions, and insufficient resource recovery [26]. To overcome such limitations, biological and/or physicochemical treatment methods are often coupled with membrane technologies. This integration delivers notable benefits in terms of efficiency and economic viability [27].

Membranes can be categorized according to their material (organic or inorganic) and structure (isotropic or anisotropic). Organic membranes are most often produced from synthetic polymers and are widely applied in pressure-based separation techniques such as microfiltration (MF), ultrafiltration (UF), nanofiltration (NF), and reverse osmosis (RO). Common materials include polyethylene (PE), polytetrafluoroethylene, polypropylene, and cellulose acetate. Non-organic membranes, on the other hand, are produced from ceramic, metal, zeolite, or silica-based materials and exhibit high chemical and thermal stability [28]. Isotropic membranes have uniform composition and structure. Microporous types offer high permeability, while dense ones have limited use due to low permeability. Microfiltration commonly employs microporous isotropic membranes. Anisotropic membranes, in contrast, have layered structures with a thin active separation layer supported by a thick, permeable layer [29].

A membrane process acts as a barrier between two phases and selectively limits the transport of certain chemicals. Membranes separate contaminants dissolved or suspended in water thanks to their semi-permeable structure. These technologies operate based on convection, diffusion of individual molecules, pressure and concentration differences, or electrical potential differences [30]. Transport mechanisms in membranes are classified as equilibrium-driven and non-equilibrium -driven processes [1, 29]. These processes are further classified as pressure-driven and non-pressure-driven membrane processes (**Figure 2**).

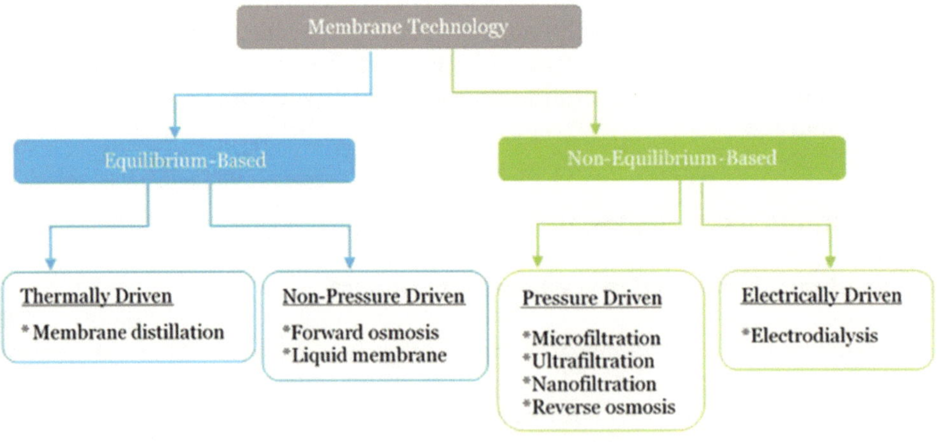

Figure 2.
Schematic diagram of membrane processes, adapted from [1].

2.1.1 Types of pressure-operated membranes

Pressure-driven membrane filtration technology is widely used to remove various pollutants from wastewater. These include heavy metal ions, oil-based pollutants, dyes, total suspended solids (TSS), microplastics, and organic and inorganic substances. The membrane types include MF, UF, NF, and RO. The fundamental difference between these membrane types is the pore size. Pore size, together with pressure levels, determines which particles or molecules can be effectively removed [2, 31]. These processes are used in many stages, from pre-treatment to final treatment. The operating principle is based on forcing water through a semi-permeable membrane by applying external pressure. During this process, contaminants are retained by the membrane, while clean water passes through. Microfiltration (0.1–10 μm) removes particles and microorganisms. Ultrafiltration (0.01–0.05 μm) removes proteins and some viruses. Nanofiltration (<0.002 μm) removes small organic molecules, hardness ions, and some salts, while reverse osmosis ($<5 \times 10^{-4}$ μm) removes dissolved salts and ions (**Table 1**).

Membrane-based methods, such as NF and RO, are classified as high-pressure membrane processes. By contrast, ultrafiltration and microfiltration are considered low-pressure membrane processes [33]. MF membranes typically operate at low differential pressures of 1–3 bar. Although these membranes are used to remove suspended solids and colloidal dyes, they cannot completely prevent the passage of dissolved organic pollutants or certain chemicals [34]. Ultrafiltration also operates at low pressure (2–5 bar), similar to microfiltration, and is used to remove viruses, bacteria, and microplastics ranging in size from 0.01 to 0.05 μm. The use of MF and UF processes as pre-treatment before NF and, more commonly, RO processes is becoming increasingly widespread today. The main purpose of this application is to reduce fouling that may occur in RO membranes and to ensure that the constant flux value is maintained. These processes are also applied in final treatment stages for organic contaminant removal and pH adjustment. They are often combined with chemical precipitation to remove phosphorus, hardness, and metals [29, 33].

The widespread use of nanofiltration membranes in wastewater treatment is attributed to their fine pore size (<0.002 μm) and superior selectivity under

Membrane type	Typical pore size (μm)	Operating pressure (Pa)	Targeted contaminants	Application areas
Microfiltration (MF)	0.1–10	10^3 - $3*10^3$	Suspended solids, some bacteria, and large particles	Pretreatment, drinking water and wastewater treatment, food industry
Ultrafiltration (UF)	0.01–0.05	$2*10^3$ – $5*10^3$	Viruses, proteins, oils, microplastics	Wastewater treatment, membrane bioreactors, dairy, and beverage industry
Nanofiltration (NF)	<0.002	$5*10^3$ – $15*10^3$	Divalent ions, pesticides, some organic compounds, dyes, and dyeing	Industrial wastewater, paint waste, drinking water treatment
Reverse Osmosis (RO)	$<5*10^{-4}$	$15*10^3$ – $75*10^3$	Monovalent ions (Na^+, Cl^-), dissolved salts, pathogens	Seawater desalination, reuse, high-purity water production

Table 1.
Categorization of pressure-assisted membrane systems applied in wastewater treatment, adapted from [29, 32].

high-pressure conditions [35]. In NF processes, separation depends not only on molecular size but also on electrical charge. Due to their negatively charged structures, they more effectively remove monovalent anions while retaining divalent anions (e.g., sulfate) to a greater extent [36]. These membranes are commonly used for removing multivalent cations and reducing disinfection by-product precursors such as color, tannin, turbidity, and organic matter. RO is an effective membrane technology for treating industrial wastewater into clean water. It can also convert wastewater into recoverable resources for direct environmental discharge or reuse for different purposes [37]. Various traditional RO module types have been developed, such as plate and frame, tubular, hollow fiber, and spiral-wound modules [38]. RO technology stands out as the most widely preferred desalination method globally due to its low unit cost advantage [39].

2.1.2 Forward osmosis

Forward osmosis (FO) is an osmotic-driven separation method. It uses a semi-permeable FO membrane to extract water from dissolved solutes. As shown in **Figure 3**, the feed solution (FS) is located on one side of the membrane, while the opposite side contains a draw solution with a high salt concentration [39]. The highly concentrated draw solution (DS) creates the necessary concentration gradient to draw water molecules from the FS. This gradient creates an osmotic pressure difference that drives water from the FS to the DS. Water continues to move through the membrane until chemical potential equilibrium is achieved [29]. Thus, the concentrated DS creates an osmotic pressure that allows water to pass through the membrane. During this process, most contaminants and salts are retained, and the water is then separated from the diluted DS [40].

In the FO process, unlike other membrane technologies, the final product is not purified water but a diluted feed solution. Therefore, an additional separation step is required to make the feed solution reusable and obtain purified water. However, this regeneration step leads to additional energy consumption and, in some cases, increases the total energy cost [41]. Therefore, one of the most important factors in the development of FO membranes is the selection of a suitable draw solution (DS). An ideal DS should exhibit high osmotic pressure, low reverse solvent

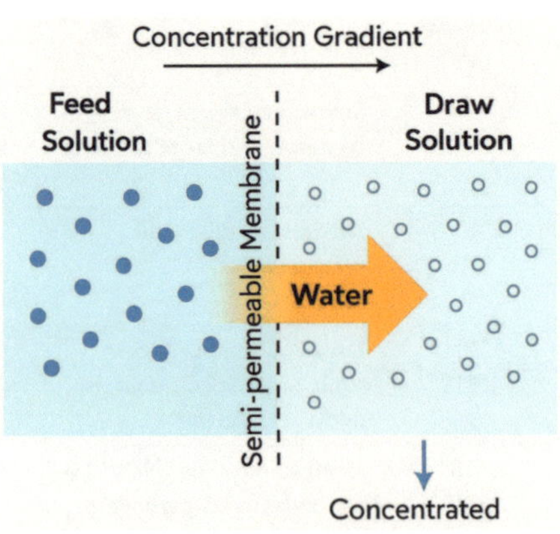

Figure 3.
Schematic representation of the forward osmosis system, modified from [29].

diffusion, easy recoverability, non-toxicity, high chemical stability, high solubility in water, and low precipitation tendency [29]. Due to its high solubility and ease of reconcentration using classical desalination methods, NaCl is one of the most commonly used attractant solutions. In recent years, studies on the selection of suitable DS have focused on inorganic (NaCl, $NaNO_3$, $MgSO_4$, $Al_2(SO_4)_3$, $CuSO_4$) and organic options (glucose, fructose, micelle solution, oxalic acid complexes containing Fe/Cr/Na, trimethylamine–carbon dioxide, CO_2-sensitive polymers such as PDMAEMA, and dimethyl ether). In addition, functionalized nanoparticle-based (e.g., magnetic nanoparticles, Na^+-functionalized carbon quantum dots) alternatives have been proposed [40].

2.1.3 Ion exchange membrane

Based on their charge groups, ion exchange membranes are categorized as anion exchange membranes (AEMs) or cation exchange membranes (CEMs). AEMs contain fixed positive groups in the polymer structure, while CEMs contain fixed negative groups. In these membranes, counterions are retained, while anions or cations in the effluent pass through [33]. The main technologies based on this principle are electrodialysis (ED), reverse electrodialysis (RED), diffusion dialysis (DD), and the Donnan membrane process (DMP). These methods rely on the selective transport of ions through membranes under an electrical potential. In the ED process shown in **Figure 4**, the AEM allows only anions to pass while retaining cations; the CEM, on the other hand, allows only cations to pass while retaining anions [29]. The applied current creates a potential gradient that enables the concentration or dilution of ions in an aqueous solution through ion migration [42].

ED and RED systems consist of four basic components: electrodes, ion exchange membranes, solvents, and electrolytes. The electrodes enable the conversion of ions into electrons, the membranes allow selective ion transport, solvents ensure continuity for ion transport, and electrolytes act as current carriers between the anode and cathode [43]. RED involves periodically reversing the electrodes and the membranes, thereby changing the direction of ion movement. This process dilutes concentrated streams while concentrating diluted ones [29]. DD is used for the recovery of free mineral acids and alkalis, while DMP is used for the recovery of toxic and valuable metals. Both methods are based on the Donnan equilibrium. However,

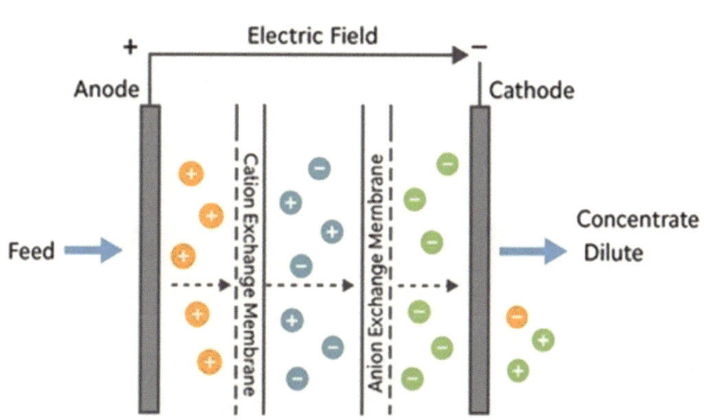

Figure 4.
Configurations of electrodialysis, modified from [33].

unlike ED and RED, DD and DMP are still being developed on a laboratory scale. They are being evaluated through kinetic studies for different effluents due to their osmotic limitations [33]. DD, using anion exchange membranes (AEM), is a self-sustaining process driven by concentration gradients without the need for additional energy input [44].

2.1.4 Membrane distillation

Membrane distillation (MD) has developed into a promising technique capable of addressing the limitations associated with pressure-driven membrane processes [45]. MD is a thermal separation technique driven by a thermal gradient, where only vapor-phase molecules pass through a porous, hydrophobic membrane [46]. In MD, mass transfer is driven by the vapor pressure gradient established through a temperature difference across the membrane. This technique offers several advantages: it operates at relatively low temperature and pressure, is cost-efficient, rejects salts and contaminants effectively, is compatible with hybrid separation systems, and shows reduced fouling tendencies owing to its porous nature [47]. This process is particularly effective for separating feed solutions with high water content. It has four main types, classified by mass transfer mechanisms on the permeate side: direct contact (DCMD), air gap (AGMD), sweep gas (SGMD), and vacuum-assisted membrane distillation (VAMD).

DCMD is currently the most widely used MD configuration in terms of applicability and scalability; however, different membrane distillation configurations are also proposed in the literature. One surface of the membrane is in direct contact with the feed solution, while the other surface is in direct contact with the cooled permeate stream. The temperature difference creates a vapor pressure gradient. As a result, water vapor passes through the membrane pores and condenses [48]. VAMD accelerates the passage of vapor through the membrane pores by applying vacuum on the permeate side. Condensation occurs outside the membrane, typically in a condenser. This method is effective for separating volatile compounds [46]. In AGMD, an air layer separates the membrane from the condensation surface. Water vapor diffuses across this gap and condenses upon reaching the cooled surface. The process is considered energy-efficient because it minimizes heat loss [47]. SGMD involves transporting vapor by supplying an inert gas flow to the permeate side and condensing it in a condenser outside the membrane. Operationally, SGMD is a promising method due to its high mass transfer coefficient and relatively low conductive heat loss. Indeed, it has been reported that SGMD provides higher permeate flux compared to AGMD and DCMD [49].

3. Application of membrane technologies in wastewater treatment: Advantages and limitations

Membrane technologies were first integrated into wastewater treatment systems during the 1960s, following the emergence of membrane bioreactor (MBR) applications [50]. An MBR is a treatment configuration that integrates membrane processes, typically ultrafiltration or microfiltration, with suspended-growth bioreactor systems [51]. These systems combine physical membrane filtration with biological treatment, ensuring the effective removal of suspended solids and microorganisms. Since the 1970s, advances in membrane materials, pore sizes, and module designs have enabled MBRs to become more compact and efficient systems. Since the 1980s, they have been widely used in water reclamation, seawater desalination, and industrial wastewater treatment [2]. Today, membrane technologies are applied in numerous areas, including water purification, wastewater management, industrial processes, food production, and biotechnology. In this context, dynamic membranes (DM), gravity-driven membrane systems (GD) [52], graphene oxide (GO) [53], and polymeric and ceramic-based membranes [54–56] have gained prominence. DM and GD membrane technologies offer energy savings due to low operating pressure and the elimination of pump requirements. Their low investment costs make them an attractive alternative for water purification and microplastic removal, particularly in developing countries[52].

Membrane technologies in wastewater treatment are increasingly preferred due to their high treatment efficiency, compact design, and ability to provide advanced water quality. Membrane processes used in domestic and industrial wastewater treatment can be utilized alone or in combination with chemical and microbial treatment methods. They can also be applied in advanced stages of treatment [2]. Membrane technologies are effectively employed in the treatment of high organic load wastewater from the paper and food industries, protein and fat-containing waste from poultry slaughterhouses, oily industrial wastewater, textile-derived waste, and industrial discharges containing heavy metals [7, 15]. MBR and membrane technologies provide high treatment efficiency in the removal of new-generation micro-pollutants of organic and inorganic origin. The literature indicates that pesticides, antibiotics, analgesics, microplastics, various industrial chemicals, phthalates, pharmaceuticals and personal care products (PCPs), surfactants, plasticizers, flame retardants, and nanomaterials have been effectively removed by MBR systems [33, 57, 58].

Membrane technologies are used not only in the treatment of domestic and industrial wastewater but also in the recovery of water from wastewater. These technologies, when applied alone or in combination with hybrid treatment processes, can ensure that the resulting water exhibits properties close to drinking water quality [4]. Membrane separation technologies are also used to recover essential elements, including lactose and protein. This provides economic benefits to the food, dairy, and pharmaceutical industries [6]. Membrane processes, as exemplified by MF, UF, NF, and RO, are widely used to obtain protein and lactose concentrates from whey and second whey, as well as to recover water from whey [36, 59–61].

Membrane technology stands out in water treatment due to its advantages, such as environmentally friendly operation, relatively low cost, and simple process conditions. However, selecting the appropriate membrane is critical for sustainable and efficient applications [62]. Modules used in membrane processes play a critical role in protecting the membranes, regulating flow, and increasing process efficiency. Different module types, like spiral-wound, tubular, plate, and hollow fiber (**Figure** 5), are preferred according to the requirements of the application. Module selection depends on the properties of the fluid, operating pressure, ease of cleaning, cost, and the targeted separation performance [29]. Correct module selection is an important step that directly affects both process efficiency and membrane lifespan.

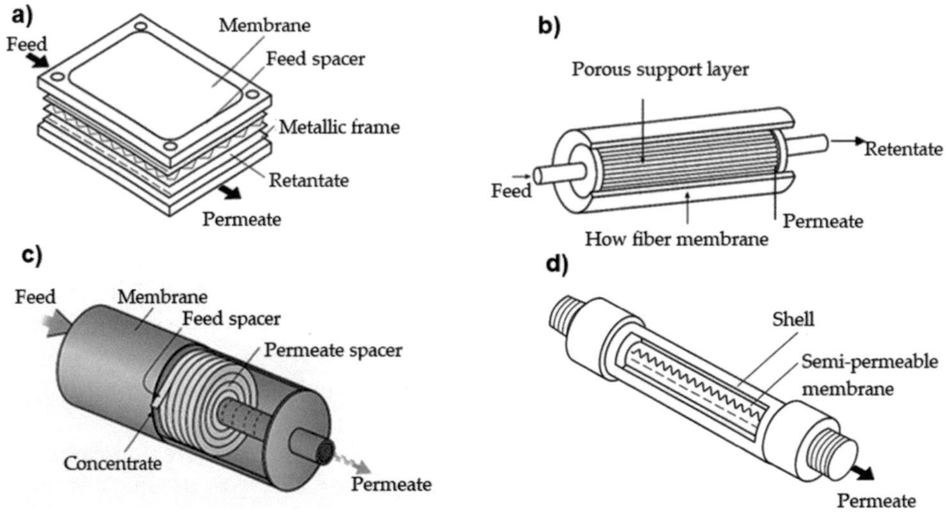

Figure 5.
Membrane module types: (a) Plate and frame membrane module, (b) Hollow fiber membrane module, (c) Spiral wound membrane module, (d) Tubular membrane module, adapted from [29, 63].

A significant advantage of MBR used for treating a wide variety of wastewater is that no coagulant needs to be added for the filtration process. This encourages the reuse of filtered water and facilitates the separation of algal biomass [64]. Additionally, they offer advantages such as small space requirements, simple transport configuration, and the ability to handle high biomass concentrations [33]. Currently, two different MBR configurations are widely used. Cross-flow MBR systems offer easy access to the membrane and cleaning advantages. In contrast, the submerged MBR system introduced by Yamamoto et al. has become widespread due to its lower energy requirements and simplicity [29]. Pressure-driven membrane processes offer advantages including high separation efficiency, low space requirements due to compact design, and energy efficiency. Additionally, their low chemical usage, suitability for continuous and automatic operation, and broad applicability against various pollutants are significant advantages [22, 65]. Reverse osmosis is the most widely used method for desalinating seawater and brine for industrial purposes via a semi-permeable membrane due to its low unit cost [66, 67]. Furthermore, its integration with solar energy systems provides a significant advantage in terms of both performance improvement and environmental and economic sustainability [39]. FO technology offers low energy consumption, simple structure and operation, and low fouling tendency. It also provides a high rejection rate across a wide range of contaminants and reduced irreversible fouling due to its operation at ambient temperature [33]. ED and RED technologies effectively remove total dissolved solids (TDS) and ionized components from wastewater. High water recovery, low pretreatment requirements, reduced membrane fouling with counter electrodes, and compatibility with renewable energy are their main advantages [29, 68]. Membrane distillation is suitable for waste heat and renewable energy integration. It operates at low temperatures and produces high-purity products by completely removing non-volatile compounds [69]. Diffusion distillation demonstrates significant economic and technical advantages over other traditional methods, including neutralization, precipitation, distillation, and solvent extraction [44].

The main limitations of MBR and other membrane technologies in wastewater treatment are clogging and the management of biofilm thickness. These factors directly determine the performance efficiency of the method [65]. Membranes have a limited lifespan due to clogging or fouling. Therefore, the need to regularly replace membranes makes this process economically unfeasible [70]. In MF membranes, fouling blocks pores with organic and inorganic deposits. This reduces permeability, selectivity, and lifespan. Although MF membranes are necessary in RO pretreatment, they pose serious problems in treating biologically contaminated wastewater [71]. A key disadvantage of MBR is that it only performs separation. It does not fully remove emerging contaminants (ECs) but only transfers them to another phase. The concentrated phase produced must be treated further before discharge [57]. Due to its high energy and pressure requirements, cross-flow MBR is less preferred today. In submerged MBR, the presence of membrane units inside the tank complicates cleaning procedures [29]. RO technology provides limited water recovery in a single module due to its pressure-based operation. Multi-stage systems are needed for high recovery, which increases energy use [66]. Reverse osmosis also has several disadvantages. These include high initial and operating costs, excessive wastewater production, membrane fouling, removal of essential minerals, limited application area, high pressure requirements, and slow water production [67]. Electrodialysis membranes consume more energy at high salinity levels, which raises costs. Additional treatment is also required because non-ionized components and microorganisms remain. Moreover, chlorine gas formed in the anode region can cause corrosion [29].

4. New developments and future perspectives

Membrane modification involves adding materials, making structural changes, or using innovative production methods to enhance membrane selectivity, durability, and performance. Its main goal is to improve contaminant separation and enable more efficient, environmentally friendly water treatment [7]. Hybrid membrane processes (HMP) combine biological and/or physicochemical treatment methods with membrane processes. They offer advantages in separation efficiency and cost. These configurations integrate membrane-based filtration with techniques such as ion exchange, adsorption, catalytic ozonation, photocatalysis, coagulation, electrocoagulation, and humidification-dehumidification [27]. Hybrid systems integrate physical, chemical, and biological processes. They enhance performance and sustainability by combining the advantages of each method while overcoming their limitations [22]. The main advantages of hybrid treatment are that it is a stable, effective, sustainable, and energy-efficient method. However, high maintenance, operating, and membrane replacement costs are among the disadvantages of this method [72]. For example, Dufton and colleagues showed that pulsed electric fields (PEF) applied to electrodialysis increase efficiency by reducing sedimentation. PEF was also effective in producing dryable acid whey [73]. A hybrid system combining reverse osmosis and membrane distillation, applied at the water/oil separator outlet, removed over 93% of organic, inorganic, and oil-based contaminants.

It also produced high-quality permeate suitable for reuse [39]. The combined use of pressure-driven membrane processes (UF–RO, MF–RO) achieves high removal rates of COD, dissolved solids, metals, and color in wastewater [72]. The FO–RO combination, on the other hand, enables effective seawater desalination [29].

In membrane technologies, carbon nanotube and graphene-based materials are attracting attention due to their high permeability, ease of cleaning, and reuse. However, limitations such as mechanical and thermal stability issues and membrane fouling may be observed [74]. Integrating carbon nanomaterials (CNM) into membranes improves properties such as durability, hydrophilicity, water permeability, and antifouling. CNT and graphene oxide stand out with their high surface area and thermal–chemical stability. In hybrid and thin-film nanocomposite membranes, they enable effective removal of water/oil emulsions and heavy metals. Furthermore, biodegradable materials such as chitosan and GO offer advantages due to their low cost and environmental compatibility. However, limitations such as brittle structure, insufficient dispersion, and weak surface interactions [1], necessitate further development of these technologies. Recently, several new materials have been developed in addition to CNT and GO. These include MOFs (metal-organic frameworks with high porosity), COFs (covalent organic frameworks), electro-curved fiber nanomembranes, MXenes (2D structure and conductivity), PDA coatings (antifouling properties), and aquaporin-based membranes. Nanoparticles such as $AgTiO_2$, SiO_2, and ZnO are also being integrated into these systems [74–77]. These new-generation materials enhance efficiency and sustainability in membrane technologies with their high permeability, selectivity, durability, and antifouling properties.

Recently, dynamic membranes have emerged as a new technology in wastewater treatment. This approach relies on a "cake layer" formed by retained microplastics and other particles on a supporting membrane, where the layer itself provides the filtration resistance [65]. Wastewater treatment plant sludge has been identified as an effective alternative for dynamic membranes under high-flow conditions. It offers cost-effective and reliable filtration on wide-pore nets [78]. Dynamic membranes have also shown very high dye removal efficiency from synthetic textile wastewater, which is difficult to treat [79].

Artificial intelligence makes significant contributions to improving the efficiency of membrane bioreactors in wastewater treatment. Approaches based on artificial neural networks can achieve 95.5% efficiency in wastewater treatment and provide savings of up to 30% in operating costs [80]. Machine learning methods, particularly artificial neural networks, are emerging as a reliable alternative for modeling membrane processes and predicting flux decline [81]. In another study, a combination of three-dimensional electrochemical treatment and adsorption achieved very high removal of methyl orange dye from textile wastewater. AI-based optimization using artificial neural networks and random forest models demonstrated the highest prediction accuracy [82]. Similarly, combining AI with response surface methodology was used to optimize the electroless nickel plating (ENP) technique in polyvinylidene fluoride–nickel membrane production. This approach determined the optimal production conditions and confirmed the high flux and contaminant removal performance of the resulting membrane [83].

5. Conclusion

In wastewater management, membrane-based processes stand out as a significant alternative owing to their high treatment efficiency, compact configurations, and adaptability to diverse processes. A variety of membrane-based techniques, including microfiltration, ultrafiltration, nanofiltration, reverse osmosis, forward osmosis, ion-exchange membranes, and membrane distillation, are employed to achieve efficient treatment of both municipal and industrial wastewater. Nevertheless, challenges, including high energy and cost demands, as well as membrane fouling, remain major constraints hindering their widespread application. In this context, the advancement of hybrid processes and integrated systems offers more sustainable solutions from both technical and economic perspectives. Looking ahead, it is anticipated that innovative approaches supported by nanotechnology, novel membrane materials, and artificial intelligence-based smart monitoring systems will further enhance the efficiency and applicability of membrane technologies in wastewater treatment.

Author details

Hakan Çelikten[1,2*]

1 Kafkas University, Faculty of Engineering and Architecture, Department of Civil Engineering, Kars, Türkiye

2 Kafkas University, Institute of Science, Department of Interdisciplinary Biotechnology, Kars, Türkiye

*Address all correspondence to: hakancelikten@kafkas.edu.tr

References

[1] Chadha U, Selvaraj SK, Vishak Thanu S, Cholapadath V, Abraham AM, M ZM, et al. A review of the function of using carbon nanomaterials in membrane filtration for contaminant removal from wastewater. Materials Research Express. 2022;**9**:012003

[2] Sangamnere R, Misra T, Bherwani H, Kapley A, Kumar R. A critical review of conventional and emerging wastewater treatment technologies. Sustainable Water Resources Management. 2023;**9**(2):58

[3] Lin L, Yang H, Xu X. Effects of water pollution on human health and disease heterogeneity: A review. Frontiers in Environmental Science. 2022;**10**:880246

[4] Hernández K, Muro C, Ortega RE, Velazquez S, Riera F. Water recovery by treatment of food industry wastewater using membrane processes. Environmental Technology. 2021;**42**(5):775–788

[5] Rashid M, Manzoor M, Mukhtar S. Urbanization and its effects on water resources: An exploratory analysis. Asian Journal of Water, Environment and Pollution. 2018;**15**:67–74

[6] Mansor ES, Ali EA, Shaban AM. Tight ultrafiltration polyethersulfone membrane for cheese whey wastewater treatment. Chemical Engineering Journal. 2021;**407**:127175

[7] Tesalonika L, Manurung TW. Studi literatur: Aplikasi teknologi membran dalam pengolahan limbah cair untuk keberlanjutan lingkungan. Prosiding Seminar Nasional Sains Dan Teknologi "Saintek". 2024;**1**(2):561–574

[8] Damania R, Desbureaux S, Rodella AS, Russ J, Zaveri E. Quality Unknown: The Invisible Water Crisis. Washington, DC: World Bank Publications; 2019. p. 216

[9] Bochynska S, Duszewska A, Maciejewska-Jeske M, Wrona M, Szeliga A, Budzik M, et al. The impact of water pollution on the health of older people. Maturitas. 2024;**185**:107981

[10] Malik OA, Hsu A, Johnson LA, De Sherbinin A. A global indicator of wastewater treatment to inform the Sustainable Development Goals (SDGs). Environmental Science & Policy. 2015;**48**:172–185

[11] Altowayti WAH, Shahir S, Othman N, Eisa TAE, Yafooz WMS, Al-Dhaqm A, et al. The role of conventional methods and artificial intelligence in the wastewater treatment: A comprehensive review. Processes. 2022;**10**:1832

[12] Iqbal J, Howari FM, Mohamed AMO, Paleologos EK. Assessment of radiation pollution from nuclear power plants. In: Mohamed AMO, Paleologos EK, Howari FM, editors. Pollution Assessment for Sustainable Practices in Applied Sciences and Engineering. Amsterdam: Butterworth-Heinemann; 2021. p. 1027–1053

[13] Dąbrowski W, Żyłka R, Malinowski P. Evaluation of energy consumption during aerobic sewage sludge treatment in dairy wastewater treatment plant. Environmental Research. 2017;**153**:135–139

[14] Tatoulis TI, Tekerlekopoulou AG, Akratos CS, Pavlou S, Vayenas DV. Aerobic biological treatment of second cheese whey in suspended and attached growth reactors. Journal of Chemical Technology & Biotechnology. 2015;**90**(11):2040–2049

[15] Yurtsever A, Sahinkaya E, Ö A, Uçar D, Ö Ç, Wang Z. Performances of anaerobic and aerobic membrane bioreactors for the treatment of synthetic textile wastewater. Bioresource Technology. 2015;**192**:564–573

[16] Luján-Facundo MJ, Mendoza-Roca JA, Cuartas-Uribe B, Álvarez-Blanco S. Membrane fouling in whey processing and subsequent cleaning with ultrasounds for a more sustainable process. Journal of Cleaner Production. 2017;**143**:804–813

[17] Ghanim MS, Soydemir G, Yılmaz F, Dizge N, Perendeci NA, Karagunduz A. Comparative study of anaerobic dynamic membrane bioreactor and anaerobic electro-dynamic membrane bioreactor systems for high-strength wastewater treatment. International Journal of Environmental Science and Technology. 2025;**22**:1–14

[18] Eniola JO, Kumar R, Barakat MA, Rashid J. A review on conventional and advanced hybrid technologies for pharmaceutical wastewater treatment. Journal of Cleaner Production. 2022;**356**:131826

[19] Crini G, Lichtfouse E. Advantages and disadvantages of techniques used for wastewater treatment. Environmental Chemistry Letters. 2019;**17**(1):145–155

[20] Mejía-Marchena R, Maturana-Córdoba A, Gómez-Cerón D, Quintero-Monroy C, Arismendy-Montes L, Cárdenas-Pérez C. Industrial wastewater treatment technologies for reuse, recycle, and recovery: Advantages, disadvantages, and gaps. Environmental Technology Reviews. 2023;**12**(1):205–250

[21] Osman AI, Chen Z, Elgarahy AM, Farghali M, Mohamed IMA, k PA, et al. Membrane technology for energy saving: Principles, techniques, applications, challenges, and prospects. Advanced Energy and Sustainability Research. 2024;**5**:2400011

[22] Khader EH, Mohammed TJ, Albayati TM, Harharah HN, Amari A, Saady NMC, et al. Current trends for wastewater treatment technologies with typical configurations of photocatalytic membrane reactor hybrid systems: A review. Chemical Engineering and Processing - Process Intensification. 2023;**192**:109503

[23] van Lier JB. High-rate anaerobic wastewater treatment: Diversifying from end-of-the-pipe treatment to resource-oriented conversion techniques. Water Science and Technology. 2008;**57**(8):1137–1148

[24] Smith VH, Schindler DW. Eutrophication science: Where do we go from here? Trends in Ecology & Evolution. 2009;**24**(4):201–207

[25] Vergara-López S, Domínguez MC, Conejo MC, Pascual Á, Rodríguez-Baño J. Wastewater drainage system as an occult reservoir in a protracted clonal outbreak due to metallo-β-lactamase-producing Klebsiella oxytoca. Clinical Microbiology and Infection. 2013;**19**(11):E490–8

[26] Zhang X, Zhang M, Liu H, Gu J, Liu Y. Environmental sustainability:

A pressing challenge to biological sewage treatment processes. Current Opinion in Environmental Science & Health. 2019;**12**:1–5

[27] Molinari R, Lavorato C, Argurio P. Application of hybrid membrane processes coupling separation and biological or chemical reaction in advanced wastewater treatment. Membranes. 2020;**10**(10):281–281

[28] Aliyu UM, Rathilal S, Isa YM. Membrane desalination technologies in water treatment: A review. Water Practice and Technology. 2018;**13**(4):738–752

[29] Obotey Ezugbe E, Rathilal S. Membrane technologies in wastewater treatment: A review. Membranes. 2020;**10**(5):89

[30] Takht Ravanchi M, Kaghazchi T, Kargari A. Application of membrane separation processes in petrochemical industry: A review. Desalination. 2009;**235**(1–3):199–244

[31] Chollom MN. Treatment and Reuse of Reactive Dye Effluent from Textile Industry Using Membrane Technology [Master's thesis]. Durban (South Africa): Durban University of Technology, Faculty of Engineering and the Built Environment, Department of Chemical Engineering; 2014. Available from: https://web.archive.org/web/20200319053452id_/https://openscholar.dut.ac.za/bitstream/10321/1388/1/CHOLLOM_2014.pdf [Accessed: 2025-May-10]

[32] Singh R, Hankins N. Emerging Membrane Technology for Sustainable Water Treatment. Amsterdam: Elsevier; 2016. p. 492

[33] Tetteh EK, Rathilal S, Chetty M, Armah EK, Asante-Sackey D, Tetteh EK, et al. Treatment of water and wastewater for reuse and energy generation-emerging technologies. In: Eyvaz M, editor. Water and Wastewater Treatment. London: IntechOpen; 2019. DOI: 10.5772/intechopen.84474

[34] Gürses A, Güneş K, Şahin E. Chapter 5 - Removal of dyes and pigments from industrial effluents. In: Sharma SK, editor. Green Chemistry and Water Remediation: Research and Applications. Amsterdam: Elsevier; 2021. p. 135–187. Advances in Green and Sustainable Chemistry. DOI: 10.1016/B978-0-12-817742-6.00005-0

[35] Bai L. Nanofiltration membrane wastewater treatment technology for pharmaceutical enterprise wastewater. Journal of Biotech Research. 2024;**16**:264–274

[36] Talebi S, Suarez F, Chen GQ, Chen X, Bathurst K, Kentish E. A pilot study on the removal of lactic acid and minerals from acid whey using membrane technology

[37] Basile A, Cassano A, Rastogi NK, editors. Current Trends and Future Developments on (Bio-) Membranes: Reverse and Forward Osmosis: Principles, Applications, Advances. Amsterdam Kidlington, Oxford Cambridge, MA: Elsevier; 2020. p. 410

[38] Zhang Z, Wu Y, Luo L, Li G, Li Y, Hu H. Application of disk tube reverse osmosis in wastewater treatment: A review. Science of the Total Environment. 2021;**792**:148291

[39] Shehata N, Egirani D, Olabi AG, Inayat A, Abdelkareem MA, Chae KJ, et al. Membrane-based water and wastewater treatment technologies: Issues, current trends, challenges, and role in achieving

sustainable development goals, and circular economy. Chemosphere. 2023;**320**:137993

[40] Şahin D. Forward osmosis membrane technology in wastewater treatment. In: I Muharrem, I Olcay Kaplan, editors. Osmotically Driven Membrane Processes. IntechOpen; 2021. DOI: 10.5772/intechopen.97483

[41] Johnson DJ, Suwaileh WA, Mohammed AW, Hilal N. Osmotic's potential: An overview of draw solutes for forward osmosis. Desalination. 2018;**434**:100–120

[42] Ward AJ, Arola K, Thompson Brewster E, Mehta CM, Batstone DJ. Nutrient recovery from wastewater through pilot scale electrodialysis. Water Research. 2018;**135**:57–65

[43] Scialdone O, Guarisco C, Grispo S, Angelo AD, Galia A. Investigation of electrode material – Redox couple systems for reverse electrodialysis processes. Part I: Iron redox couples. Journal of Electroanalytical Chemistry. 2012;**681**:66–75

[44] Lin J, Huang J, Wang J, Yu J, You X, Lin X, et al High-performance porous anion exchange membranes for efficient acid recovery from acidic wastewater by diffusion dialysis. Journal of Membrane Science. 2021;**624**:119116

[45] Nthunya LN, Chong KC, Lai SO, Lau WJ, López-Maldonado EA, Camacho LM, et al. Progress in membrane distillation processes for dye wastewater treatment: A review. Chemosphere. 2024;**360**:142347

[46] Yan Z, Jiang Y, Liu L, Li Z, Chen X, Xia M, et al. Membrane distillation for wastewater treatment: A mini review. Water. 2021;**13**:3480

[47] Shaheen A, AlBadi S, Zhuman B, Taher H, Banat F, AlMarzooqi F. Photothermal air gap membrane distillation for the removal of heavy metal ions from wastewater. Chemical Engineering Journal. 2022;**431**:133909

[48] Gugliuzza A. Review of techniques for future manufacturing of ordered and functional porous films: Conceivable membranes for water desalination by membrane distillation processes. Advanced Manufacturing. 2024;**1**(2):1–29. DOI: 10.55092/am20240009

[49] Mousavi SA, Arab Aboosadi Z, Mansourizadeh A, Honarvar B. Surface modified porous polyetherimide hollow fiber membrane for sweeping gas membrane distillation of dyeing wastewater. Colloids and Surfaces A: Physicochemical and Engineering Aspects. 2021;**610**:125439

[50] Bera SP, Godhaniya M, Kothari C. Emerging and advanced membrane technology for wastewater treatment: A review. Journal of Basic Microbiology. 2022;**62**(3–4):245–259

[51] Shahot K, Idris A, Omar R, Mohamed Yusoff H. Review on biofilm processes for wastewater treatment. Life Science Journal. 2014;**11**:1–13

[52] Colakoglu EB, Uyanik I, Elbir H, Sahinkaya E, Yurtsever A. A novel gravity-driven dynamic membrane filtration reactor for microplastic removal from plastic recycling facility wastewater. Journal of Environmental Chemical Engineering. 2025;**13**(2):115793

[53] Akin İ, Zor E, BiNgöl H. Preparation and Characterization of

GO/Fe3O4 Doped Polymeric Composite Membranes. Neufmbd. 2023;**5**(2):38–52

[54] Goswami KP, Pugazhenthi G. Credibility of polymeric and ceramic membrane filtration in the removal of bacteria and virus from water: A review. Journal of Environmental Management. 2020;**268**:110583

[55] Wan H, Shi K, Yi Z, Ding P, Zhuang L, Mills R, et al. Removal of polystyrene nanoplastic beads using gravity-driven membrane filtration: Mechanisms and effects of water matrices. Chemical Engineering Journal. 2022;**450**:138484

[56] Kim LH, Lee D, Oh J, Kim S, Chae SH, Youn D, et al. Performance of a novel granular activated carbon and gravity-driven membrane hybrid process: Process development and removal of emerging contaminants. Process Safety and Environmental Protection. 2022;**168**:810–819

[57] Rout PR, Zhang TC, Bhunia P, Surampalli RY. Treatment technologies for emerging contaminants in wastewater treatment plants: A review. Science of the Total Environment. 2021;**753**:141990

[58] Tuan Tran H, Lin C, Bui XT, Ky Nguyen M, N DTC, Mukhtar H, et al. Phthalates in the environment: Characteristics, fate and transport, and advanced wastewater treatment technologies. Bioresource Technology. 2022;**344**:126249

[59] Valta K, Damala P, Angeli E, Antonopoulou G, Malamis D, Haralambous KJ. Current treatment technologies of cheese whey and wastewater by greek cheese manufacturing units and potential valorisation opportunities. Waste and Biomass Valorization. 2017;**8**(5): 1649–1663

[60] Pires AF, Marnotes NG, Rubio OD, Garcia AC, CD P. Dairy by-products: A review on the valorization of whey and second cheese whey. Foods. 2021; **10**(5):1067

[61] Blais HN, Schroën K, Tobin JT. A review of multistage membrane filtration approaches for enhanced efficiency during concentration and fractionation of milk and whey. International Journal of Dairy Technology. 2022;**75**(4):749–760

[62] Tanis-Kanbur MB, Peinador RI, Calvo JI, Hernández A, Chew JW. Porosimetric membrane characterization techniques: A review. Journal of Membrane Science. 2021;**619**:118750

[63] Sommer S, Klinkhammer B, Schleger M, Melin T. Performance efficiency of tubular inorganic membrane modules for pervaporation. AIChE Journal. 2005;**51**(1):162–177

[64] Younas F, Mustafa A, Farooqi ZUR, Wang X, Younas S, Mohy-Ud-Din W, et al. Current and emerging adsorbent technologies for wastewater treatment: Trends, limitations, and environmental implications. Water. 2021;**13**:215

[65] Krishnan RY, Manikandan S, Subbaiya R, Karmegam N, Kim W, Govarthanan M. Recent approaches and advanced wastewater treatment technologies for mitigating emerging microplastics contamination – A critical review. Science of the Total Environment. 2023;**858**:159681

[66] Judd SJ. Membrane technology costs and me. Water Research. 2017;**122**:1–9

[67] Faroon M, ALSaad Z, Albadran F, Ahmed L. Review on technology-based on reverse osmosis. Anbar Journal of Engineering Sciences. 2023;**14**(1):89–97

[68] Gally CR, Benvenuti T, Da Trindade CDM, Rodrigues MAS, Zoppas-Ferreira J, Pérez-Herranz V, et al. Electrodialysis for the tertiary treatment of municipal wastewater: Efficiency of ion removal and ageing of ion exchange membranes. Journal of Environmental Chemical Engineering. 2018;**6**:5855–5869

[69] Carmona B, Abejón R. Innovative membrane technologies for the treatment of wastewater polluted with heavy metals: Perspective of the potential of electrodialysis, membrane distillation, and forward osmosis from a bibliometric analysis. Membranes. 2023;**13**(4):385

[70] Khan MD, Singh A, Khan MZ, Tabraiz S, Sheikh J. Current perspectives, recent advancements, and efficiencies of various dye-containing wastewater treatment technologies. Journal of Water Process Engineering. 2023;**53**:103579

[71] Anis SF, Hashaikeh R, Hilal N. Microfiltration membrane processes: A review of research trends over the past decade. Journal of Water Process Engineering. 2019;**32**:100941

[72] Solayman HM, MdA H, Abd Aziz A, Yahya NY, Leong KH, Sim LC, et al. Performance evaluation of dye wastewater treatment technologies: A review. Journal of Environmental Chemical Engineering 2023;**11**:109610

[73] Dufton G, Mikhaylin S, Gaaloul S, Bazinet L. Systematic study of the impact of pulsed electric field parameters (Pulse/Pause Duration and Frequency) on ED performances during acid whey treatment. Membranes. 2020;**10**(1):14

[74] Singha I, Kumar Mishrab P. Nano-membrane filtration a novel application of nanotechnology for waste water treatment. Materials Today: Proceedings. 2020;**29**:327–332

[75] Meng B, Liu G, Mao Y, Liang F, Liu G, Jin W. Fabrication of surface-charged MXene membrane and its application for water desalination. Journal of Membrane Science. 2021;**623**:119076

[76] Taş CE. Hybrid Polymeric Materials Comprising Clay Nanotubes, Photothermal Agents and Phase Change Materials for Food. Water and Energy Applications (Doctoral dissertation). 2023

[77] Asif MB, Kim S, Nguyen TS, Mahmood J, Yavuz CT. covalent organic framework membranes and water treatment. Journal of the American Chemical Society. 2024;**146**(6): 3567–3584

[78] Elbir H, İ U, Çolakoglu EB, Sahinkaya E, Yurtsever A. Pre-coating of stainless-steel mesh support material with wastewater treatment plant sludges in a dynamic membrane filtration process. Journal of Water Process Engineering. 2024;**64**:105673

[79] Yurtsever A, Basaran E, Ucar D, Sahinkaya E. Self-forming dynamic membrane bioreactor for textile industry wastewater treatment. Science of the Total Environment. 2021;**751**:141572

[80] Khan SAR, Ponce P, Yu Z, Golpîra H, Mathew M. Environmental technology and wastewater treatment: Strategies to achieve environmental sustainability. Journal of the American Chemical Society. 2022;**286**:131532

[81] Martini S, Roni KA. The existing technology and the application of digital artificial intelligent in the wastewater treatment area: A review paper. Journal of Physics: Conference Series. 2021;**1858** (1):012013

[82] Ganthavee V, Fernando MMR, Trzcinski AP. Monte Carlo simulation, artificial intelligence and machine learning-based modelling and optimization of three-dimensional electrochemical treatment of xenobiotic dye wastewater. Environmental Processes. 2024;**11**(3):41

[83] Li B, Yue R, Shen L, Chen C, Li R, Xu Y, et al. A novel method integrating response surface method with artificial neural network to optimize membrane fabrication for wastewater treatment. Journal of Cleaner Production. 2022;**376**:134236

Section 3

Advanced Wastewater Treatment

Chapter 3

Application of Nonthermal Plasma System for Wastewater Treatment

Süheyla Tongur, Şule Büşra Zambak and Sevil Yıldız

Abstract

Nonthermal atmospheric pressure plasma jets are becoming increasingly important in wastewater treatment applications. Especially, it is quite difficult to remove the color found in the wastewater of some industries such as textile and paint industries by traditional methods. Therefore, in the study, the color removal efficiency was investigated using atmospheric pressure plasma system. In the design of the system, an electric current is directed from a low pressure gas flow and an atmospheric plasma field is created. Treatment is achieved as a result of contact with the wastewater. In this study, the removal efficiencies of pomegranate wastewater and methyl orange (MO) as a model dye were investigated. Oxygen gas was studied as feed gas. Plasma discharge flow rate was applied as 12 mg/L, and plasma duration was 5 minutes. The results were evaluated for two different wastewaters. For an initial concentration of 20 mg/L, the COD removal efficiencies of methyl orange and pomegranate wastewater (PW) were found to be 83.7 and 98.4%, respectively. The highest color removal efficiency for methyl orange was calculated as 97%. In addition, *Lepidium sativum* toxicity test was used to determine whether there was any toxic effect due to radicals formed in the system. No toxic effects from the plasma system were observed. As a result, inspired by the studies conducted worldwide on the usability of plasma technology in water treatment, the usability of the atmospheric pressure plasma system, which is a new technology in the field of water treatment, was evaluated.

Keywords: atmospheric pressure plasma, methyl orange, pomegranate wastewater, *Lepidium sativum*, degradation

1. Introduction

Water pollution is a serious problem in modern society. Recently, there has been a dramatic increase in the amount of organic pollutants (volatile organic compounds, pharmaceuticals, organic dyes, and so on) detected in water resources, and this poses a serious concern for public health [1]. Colored wastewater is not just an inherent feature of textile and dye waste; it is produced in industries such as food companies, cosmetics industries, chemical and pharmaceutical companies, paper mills, and oil industries [2]. With the rapid development of the printing and dyeing industry, chemical industry, and oil industry, a large amount of low or nonbiodegradable organic dye wastewater is discharged, causing a serious environmental problem [3, 4]. In modern industrial society, large amounts of dyes and other organic pollutants are emitted into the environment and the organic contamination degradation has become the focus of environmental remediation efforts.

The production amount of dye industries in our country is increasing day by day. Since the increase in production is parallel to the amount of wastewater generated by the facilities, the amount of wastewater generated in this area is also increasing. This type of wastewater is dark colored, has high pH, contains high organic matter, and has low biodegradability [5]. The main processes of pomegranate fruit-juice and other fruit-juice productions are usually in the packing operation where water is used for general washing and cleaning, the selection and cleaning operation where water is used mainly for washing, cleaning, and moving of the fresh fruit prior to the extraction process, juice extraction and treatment, peel treatment, and emulsion treatments. These processes require large quantities of fresh water and also produce plenty of wastewater. These wastewaters are characterized by high chemical oxygen demand (COD) contents of organic matter, a low pH, an imbalance of nutrients, color, dissolved solids, and suspended solids [6, 7]. Hence, the discharge of these types of wastewaters into urban wastewater treatment plants is not allowed, thus requiring a prior treatment. The hazard effluents produced from these industries have serious problems as they are low or nonbiodegradable. The prominent problems for dyeing wastewater treatment are color removal and degradation of organic compounds [3]. Thus, they must be treated to satisfy the water quality standards before discharging into the environment [6].

The development of a new and efficient treatment technique for the organic contaminants is necessary [8]. Various methods such as physical methods [9, 10], chemical methods [11, 12], biological methods [13], and combined methods [14–16] have been proposed. Physical methods such as adsorption [17] can be effectively separated the dye from effluents, but without complete degradation, so a secondary pollution may produce by converting dye from one phase to another. Therefore, physical and chemical methods transfer the pollutants to another phase, which usually induce by-products formation, so this methods are not effective. Biological methods are not effective for the highly colored and low degradable dyeing wastewater [18]. Therefore, the development of environmentally friendly and efficient methods for wastewater treatment receives great attention [19].

Plasma technologies: it is used as a technology that has become increasingly important in recent years in the treatment of persistent, recalcitrant, and difficult-to-remove pollutants in water, requiring less chemical use during treatment and having high removal efficiency in a short time. Plasma oxidation process, with high efficiency and without secondary pollution, is one of the most promising methods for colored wastewater treatment [19]. All plasma systems are classified into two main categories, thermal and nonthermal plasma [20]. In the thermal plasma (such as arc discharges or radio frequency (RF)), sufficient energy is inserted to make the plasma constituents in a thermal equilibrium. On the other hand, less power (such as corona discharge and dielectric barrier discharge) is used to obtain the nonthermal plasma. The nonthermal plasma produces energetic electrons that collide with another molecule such as O_2, N_2, and H_2O produces secondary species include radicals, ions, photons, and electrons [21, 22]. In the remediation of the wastewater, the organic contaminates may be dissociated by the action of the energetic electrons as well as the action of the active radicals and active molecules. Furthermore, the physical effects can also act on the organic pollutants in water. Thus, the remediation of water polluted with organic compounds may be effective and rapid owing to the high oxidative potential of active species as well as the actions of physical effects [6]. Nonthermal plasma (NTP) is plasma characterized by nonequilibrium conditions: the temperature of the bulk gas, typically near room temperature, is much lower than that of the plasma reactive species, that is, free electrons and ions, atoms and radicals, and excited molecules [23]. Nowadays, NTP has many applications, ranging from microprocessors cooling [24], to surface modification, to living tissues treatments [25], to bio-decontamination [26], and to air [27] and wastewater treatments [26, 28]. In wastewater treatment, a high voltage electrical discharge is applied above the surface of the water inducing physical and chemical phenomena in the gaseous, liquid, and interfacial media. The chemical properties of such complex systems depend on the gas to which the discharge is applied. Oxidant species common to any gas are hydroxyl radicals (OH) and hydrogen peroxide (H_2O_2), formed by the interaction of water molecules with high-energy electrons [28, 29]. In particular, OH radicals are produced by the well-known water dissociation reaction, while H_2O_2 is generally formed by OH recombination [29].

In this paper, the usability of plasma technology, an innovative technology that has gained importance in recent years, in the field of water treatment has been evaluated. Within the scope of the study, an atmospheric pressure plasma system fed with oxygen gas was established to be in contact with liquid. First of all, the removal efficiency of the plasma system was investigated for the aqueous solution of methyl orange dye, which has a stable, molecular structure and is resistant to degradation. However, to test the atmospheric pressure plasma system on colored wastewater, a pomegranate wastewater sample obtained from a fruit processing plant was used. In addition, this experimental study aims to investigate plasma systems, which are an innovative technology, and contribute to the literature with future studies on environmental engineering such as water and wastewater treatment [30].

2. Material and methods

2.1 Material

2.1.1 Reagents

In this study, powdered methyl orange (ZAG Chemical) and powdered $FeSO_4$ (Carlo Erba) powder were purchased from a private medical company and used. Chemicals were used by turning them into synthetic solutions with pure water. 1 M NaOH and 2% HCl were used for pH adjustments.

2.1.2 Apparatus and measuring materials

2.1.2.1 Hydrogen peroxide (H_2O_2) test

Supelco brand MQuant hydrogen peroxide test strips, which can measure in the range of 0.5–25 mg/L (H_2O_2), were used for measurement.

2.1.2.2 pH and conductivity

The pH meter used in the study was obtained from Necmettin Erbakan University Scientific Technology and Research Center, and temperature measurement was also carried out with this device. The conductivity device was obtained from Konya

Technical University, Faculty of Engineering and Natural Sciences, Department of Environmental Engineering, Wastewater Treatment and Research Laboratory.

2.1.2.3 UV-Vis spectrophotometer

For absorbance measurement, SHIMADZU brand UV-3600 Plus model UV-VIS-NIR Spectrophotometer, located within Necmettin Erbakan University Scientific Technology and Research Center, was used.

2.1.2.4 Lepidium sativum toxicity test

For *Lepidium sativum* toxicity analyses; 9 cm diameter glass petri dishes, 90 mm diameter Whatman 1 brand filter paper, deionized water, and cress seeds were used.

2.1.2.5 Atmospheric pressure plasma system

In the study, the OptoSense brand mini atmospheric plasma system within the Necmettin Erbakan University Scientific Technology and Research Center was used. The device consists of Radiofrequency generator was shown in **Figure 1(a)**, flexible gas distribution pipe and probe. The device has a size of 380 × 280 × 280 mm

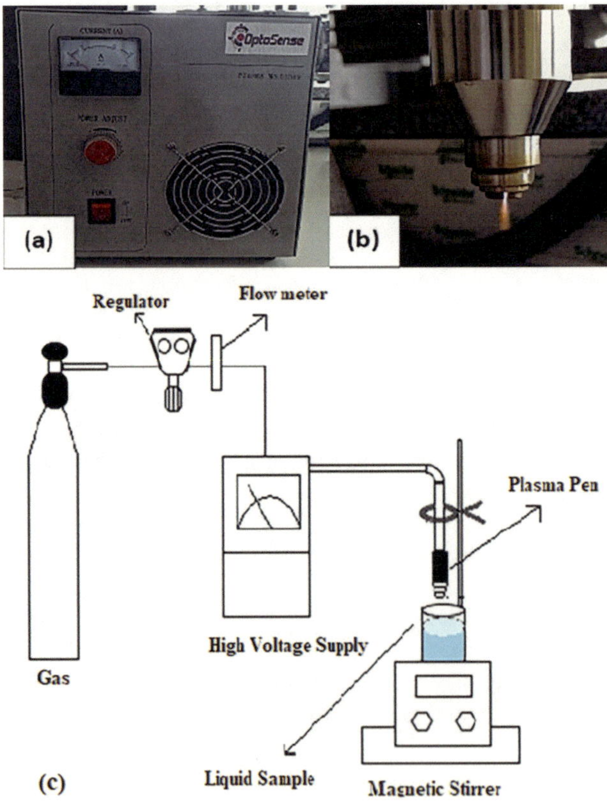

Figure 1.
(a) Plasma power supply, (b) Plasma beam, and (c) Experimental setup of atmospheric pressure plasma system [30].

UXHXW. The probe of the device has a 10–12 mm round head part and a 15–18 mm rectangular part on its body, as shown in **Figure 1(b)**. The device has 1000 VA expandable power and 220 (±5%) power. Inlet gas pressure is 2–2.5 kg 0.12 m^3/min. The device has been modified for use under laboratory conditions and made suitable for liquid sample processing. In order to increase the removal efficiency during the process, a magnetic stirrer was placed under the heat-resistant glass container containing the liquid sample. Since the device works with supply gas, it was connected to an oxygen cylinder. A flow meter was used to control the gas flow. A regulator was used to ensure pressure adjustment and gas flow. The experimental setup of the plasma system is given in **Figure 1(c)**.

For a better understanding of the work process, the workflow diagram is given in **Figure 2**.

2.2 Methods

2.2.1 Preparation of solutions

2.2.1.1 Preparation of methyl orange solution

To prepare 100 mg/L methyl orange stock solution, first 0.1 g of powdered methyl orange was taken into a 100 mL beaker and weighed on a precision balance. First, 80 mL of pure water was added to the beaker and mixed for 5 minutes in the mixer. The mixture was poured into a 1000 mL volumetric flask. By adding pure water to the

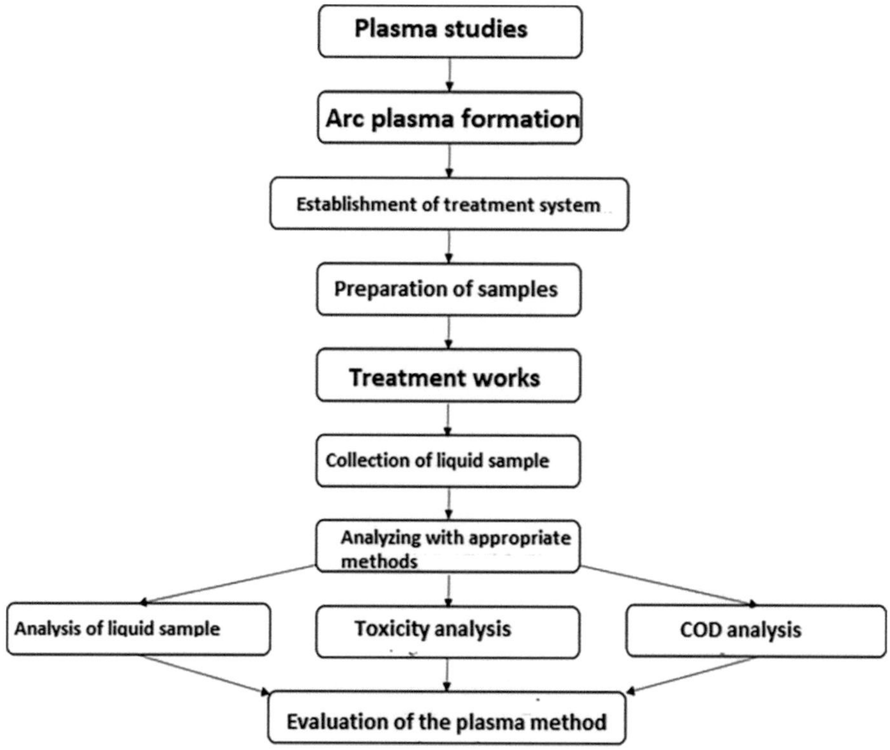

Figure 2.
Experimental flow diagram [30].

volumetric flask, it was filled to 1000 mL and shaken. 100 mg/L methyl orange stock solution was obtained. The following (Eq. (1)) was used to prepare dilute solutions at different concentrations:

$$M1 \times V1 = M2 \times V2 \tag{1}$$

$M1$ = Concentration to be prepared (mg/L); $V1$ = Volume to be prepared (mL); $M2$ = Stock solution concentration (mg/L); $V2$ = Volume to be ta ken from the stock solution (mL).

2.2.1.2 Preparation of iron 2 sulfate ($FeSO_4$) solution

To prepare a 1000 mg/L stock solution with powdered $FeSO_4$ with a molar mass of 278.05, 4.978 g of $FeSO_4$ powder was weighed and prepared according to the stock solution preparation procedure. In the study, iron II sulfate ($FeSO_4$) catalyst was used to investigate the effect of methyl orange on plasma removal efficiency.

2.2.2 Preparation of pomegranate wastewater sample

In the study, the efficiency of the atmospheric pressure plasma system was investigated in a colored wastewater sample. For this purpose, decolorization of a pomegranate wastewater sample obtained from a Fruit Processing Industry was tried with the plasma system. The supplied pomegranate wastewater sample was diluted at 100%, 25%, 10%, and 5% before the plasma study. These dilutions were taken into 50 mL sample containers for initial absorbance measurements.

2.2.3 Working principle of UV-Vis spectrophotometer

UV-Vis Spectrophotometer was used to determine color removal efficiency, and necessary calculations were made according to absorbance measurement results. For removal efficiency calculations, standard solutions of methyl orange samples were prepared at concentrations of 1, 3, 5, 7, and 10 mg/L and absorbance measurements were made. Calibration graphs were obtained with the absorbance values of the standard solutions for methyl orange at 465 nm [31]. From this calibration chart, the formula y = ax + b was obtained. When calculating the removal efficiency, the absorbance value (Co) of the initial sample was accepted as the y value in the formula. In the formula, the x value was accepted as the absorbance value (C) of the sample after plasma treatment. If dilution was made in the samples before absorbance measurement, the result was multiplied by the dilution ratio. After the C value was found, the % removal efficiency was calculated from the formula (Co-C)/Co*100 [32].

2.2.4 Lepidium sativum toxicity test procedure

Lepidium sativum acute toxicity test method was preferred for toxicity analyses. Toxicity testing was performed for the initial and post-plasma treatment samples of the methyl orange solution prepared under optimum conditions (20 mg/L concentration, 5 min. Plasma time, 5 min. Mixer time, 1.5 cm distance, 12 mg/L flow rate). *Lepidium sativum* toxicity test was performed with 6 controls and 3 methyl orange solutions prepared in dilutions of 3.125, 6.25, 12.5, 25, 50, and 100%. In the experiment, two 90 mm diameter Whatman 1 brand filter papers were placed in 9 cm glass petri dishes. 5 mL of pure water was placed in the control petri dishes, and 5 mL of the prepared sample at different dilutions was placed in the sample petri dishes, and placed so that no air bubbles remained under the filter paper. 25 undamaged *Lepidium sativum* seeds of equal size were placed in each petri dish at equal intervals, and the sealed petri dishes were incubated in a dark environment at approximately 25°C for 72 hours. At the end of the test period, the root and stem lengths of the 20 best-developing *Lepidium sativum* seeds in each petri dish were measured. At the end of the test period, the average values of root and stem lengths observed in the *Lepidium sativum* seeds of the samples were compared with the average root and stem length values measured in the control petri dishes, and the % inhibition rate and EC_{50} value were determined. The studies were evaluated according to the toxicity classification by calculating the % inhibition values and Toxic Unit value for 3.125, 6.25, 12.5, 25, 50, and 100% dilutions.

2.2.5 COD measurement

Analysis was carried out with the SM 5220 C method. For this study, initial and post-plasma COD analyzes were performed on methyl orange solution and pomegranate sample prepared under optimum conditions. % removal efficiency was calculated from the analysis results.

2.2.6 Atmospheric pressure plasma system working principle

The atmospheric pressure plasma and magnetic stirrer devices were turned on and the distance between the probe, and the sample was adjusted to the desired size. 50 mL sample was taken into the beaker where the study would be performed and placed in the magnetic stirrer in the plasma device. The magnetic stirrer was added to the system to accelerate the reactions that occur during and after the plasma process. The sample continued to be mixed for the desired periods of time while the plasma was running and after the plasma was turned off. Oxygen gas was used as feed gas in the experiments. After the gas cylinder valve was opened, the flow meter and ampere levels were fixed and the gas pressure was brought to the desired level by turning the regulator valve. The removal process was carried out by initiating the interaction of the plasma beam and the sample.

3. Results

3.1 Effect of plasma on methyl orange color removal

In this study, 3 replicate studies were carried out under the same conditions with 50 mL of methyl orange solution at a concentration of 20 mg/L to test the effect of the Atmospheric Pressure Plasma System on methyl orange removal. Absorbance measurements were read at 465 nm on a UV-Vis Spectrophotometer. Optimum conditions were provided for the study, and the removal efficiency is given in **Table 1**, and the absorbent curve graph is given in **Figure 3**.

Jiang et al. [31] used an oxygen-fed atmospheric pressure plasma system in their study. 20 min after plasma treatment, they reached 92% removal efficiency of methyl orange solution with a concentration of 40 mg/L. The results obtained were found to

Concentration (mg/L)	Initial UV (abs)	Post UV plasma (abs)	Removal efficiency (%)
20	1586	0,05	97
20	1586	0,034	98
20	1586	0,049	97

Table 1.
Effect of plasma on methyl orange removal efficiency [30].

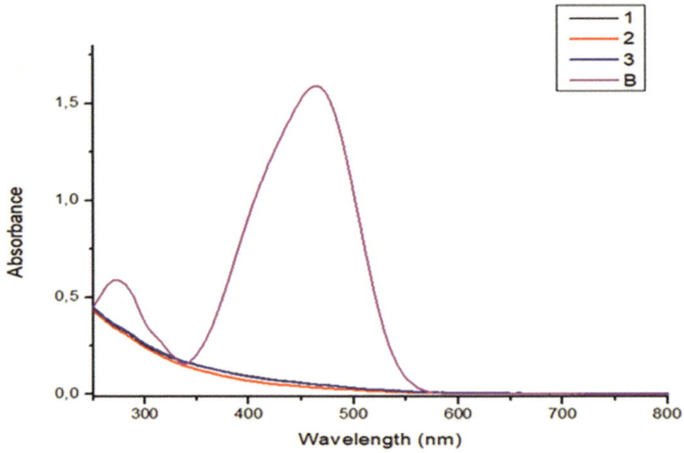

Figure 3.
Absorbance graph of methyl orange analyzes performed under the same conditions as plasma (B: Starting sample) [30].

be compatible with the literature. They stated that hydroxyl radicals and O_3, which have high oxidation potential, are effective for the degradation of pollutant color molecules, and based on the theory of bond dissociation energies, they gave suggested ways for the oxidation of MO with hydroxyl and ozone radicals in their research. Accordingly, the oxidation pathways of MO molecules and their intermediate products in the plasma treatment process were given in **Figure 4**.

In **Figure 5**, the increase-decrease slopes of pH (A) and conductivity (B) results of plasma studies performed with MO solutions were given, respectively. Accordingly, it was observed that the initial pH decreased after plasma. When the conductivity curve was examined, an increase in conductivity was observed after the plasma treatment.

3.2 Effect of catalyst

FeSO4 was preferred to examine the effect of catalyst addition on removal efficiency. For this purpose, catalyst was added at a fixed concentration of 20 mg/L to the methyl orange synthetic solution prepared at a concentration of 20 mg/L. Absorbance readings of the study were made at 465 nm for methyl orange. The removal effect of catalyst addition on methyl orange is given in **Table 2**. The absorbance curve graph is given in **Figure 6**. According to the analysis results obtained in the study, it was concluded that the reaction rate of the atmospheric pressure plasma system for methyl orange removal is high, but the addition of catalyst may not be necessary.

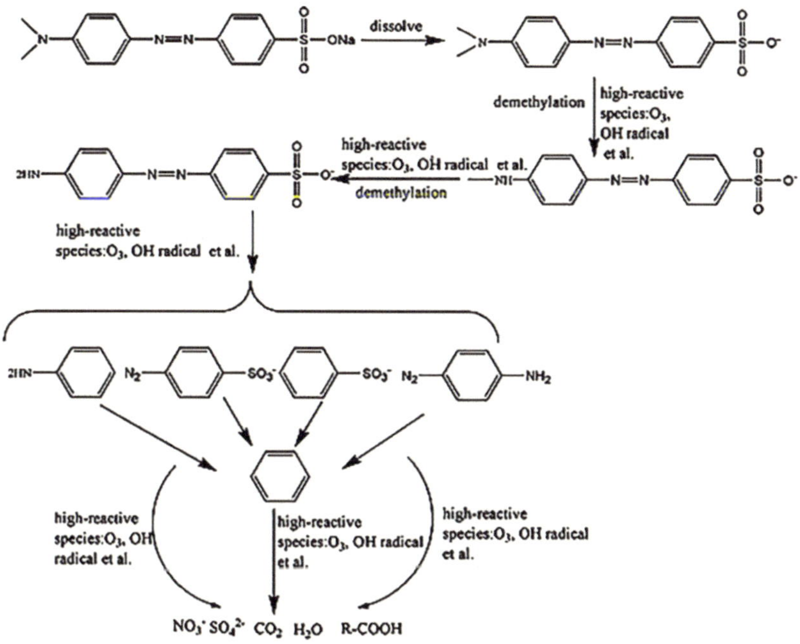

Figure 4.
Proposed pathways for mineralization of methyl orange molecules by atmospheric pressure plasma system [31].

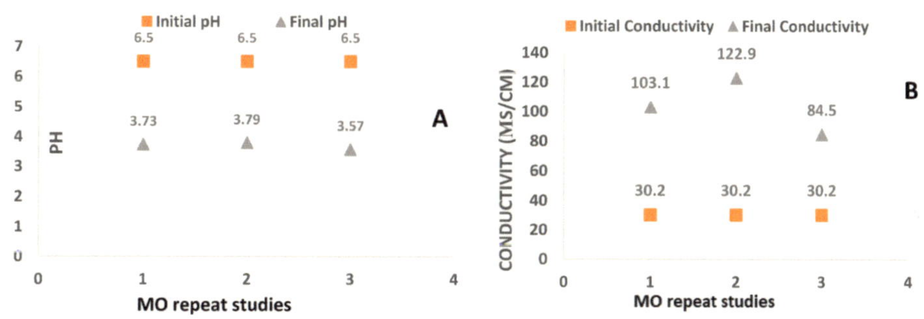

Figure 5.
Initial (before plasma) and final (after plasma) pH (A) and conductivity (B) values of 3 samples studied under the same conditions [30].

Concentration (mg/L)	Initial UV (abs)	Post UV plasma (abs)	Removal efficiency (%)
20	1582	0,142	91

Table 2.
Effect of $FeSO_4$ addition on MO removal efficiency [30].

Figure 7(A) and **(B)** shows the increase-decrease slopes of pH and conductivity results, respectively, of plasma studies conducted with catalyst-added MO solution. Studies have shown that initial pH decreases after plasma. When the conductivity curve was examined, an increase in conductivity was observed after the plasma treatment.

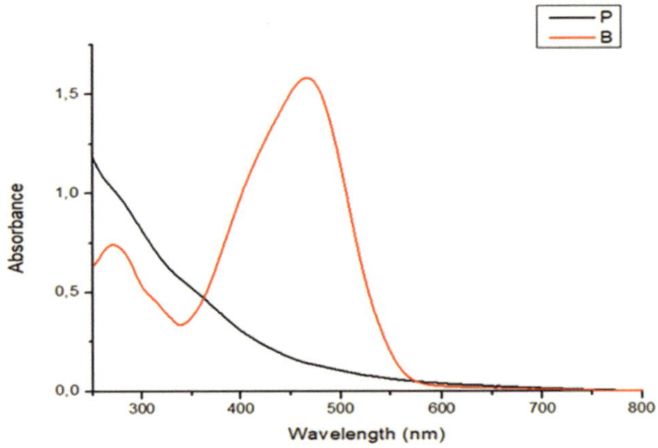

Figure 6.
Absorbance graph of samples before (B) and after (P) plasma treatment of MO solution with catalyst addition [30].

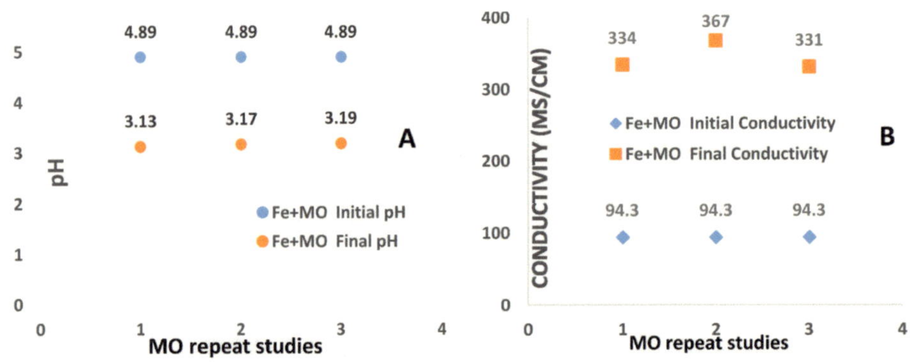

Figure 7.
Initial (pre-plasma) and final (post-plasma) pH (A) and conductivity (B) values of the catalyst-added MO sample [30].

3.3 Effect of plasma on colored wastewater sample

Oxygen gas-fed plasma study of the pomegranate wastewater sample obtained from a Fruit Processing Facility was carried out under optimum conditions. 660 nm of samples diluted in certain proportions. Absorbance values were also read. The absorbance curve graph is in **Figure 8**, and the pH and conductivity graphs are in **Figure 9(A)** and **(B)**. When looking at the absorbance values before and after plasma given in **Table 3**, a decrease in color removal was observed. According to the graphs of the increase-decrease slopes of pH and conductivity results, it was seen that the initial pH of the studies decreased after plasma. When the conductivity curve was examined, an increase in conductivity was observed after the plasma treatment.

Regarding pomegranate wastewater, reference [33] investigated the color removal of pomegranate juice with atmospheric pressure plasma in their literature study. They worked for 3, 5, 7 minutes for the plasma process and achieved color removal after the plasma process.

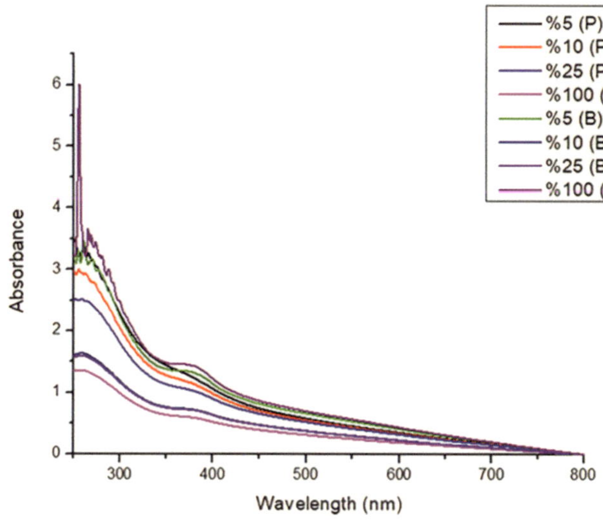

Figure 8.
Absorbance values of the pomegranate wastewater sample studied with different dilutions before (B) and after (P) plasma treatment [30].

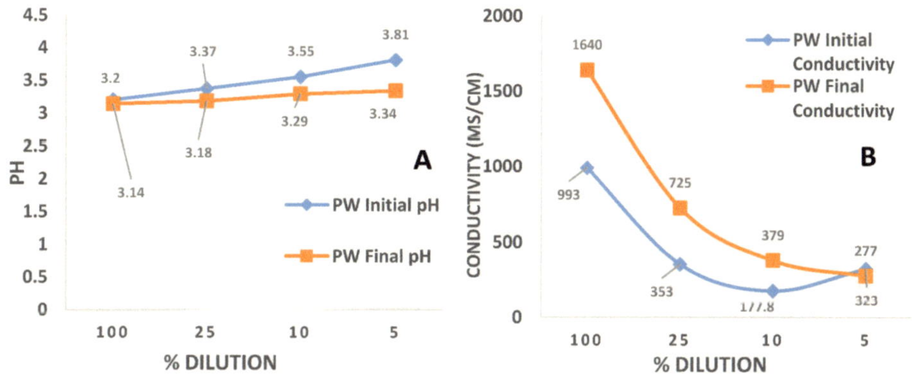

Figure 9.
Initial (before plasma) and final (after plasma) pH (A) and conductivity (B) values of the pomegranate wastewater sample studied with different % dilutions [30].

Dilution ratio (%)	Initial UV (abs)	Post UV plasma (abs)
100	0,395	0,916
25	0,284	0,299
10	0,271	0,332
5	0,152	0,238

Table 3.
Effect of plasma system on color removal of pomegranate wastewater sample [30].

3.4 COD effect of plasma

The COD pollution parameters of the methyl orange synthetic solution prepared under optimum conditions and the pomegranate wastewater sample taken from the fruit processing plant were examined before and after plasma treatment. The initial COD values, post-plasma COD values, and % removal values of the samples are given in **Table 4**, and the comparison of % removal efficiencies is given in **Figure 10**. In the literature, reference [31] used an Atmospheric pressure plasma system working with oxygen gas in their study and stated that plasma could efficiently reduce organic pollutants with a 67% COD removal efficiency as a result of a 20 minute plasma treatment for methyl orange removal.

3.5 *Lepidium sativum* toxicity test results

In the study, *Lepidium sativum* toxicity test results were determined for the methyl orange synthetic solution prepared under optimum conditions and kept in plasma discharge. In the study, dilution rates of 3.125%, 6.25%, 12.5%, 25%, 50%, and 100% were used. EC_{50} value and Toxic Unit (TU) value were calculated according to *Lepidium sativum* toxicity test results. In this study, the EC_{50} values and toxic unit classification calculated for the *Lepidium sativum* toxicity test are given in **Tables 5** and **6** for MO initial and post-plasma, respectively.

The result of the *Lepidium sativum* toxicity test was given for the initial and post-plasma solution sample according to toxic unit classification. According to the results of the study, the Toxic Unit values of the methyl orange solution were "Slightly Toxic" according to the root and hypocotyl values before and after plasma [34]. In their study, the toxic classification was found as "Toxic" according to the *Lepidium sativum*

Samples	Initial COD (mg/L)	Post plasma COD (mg/L)
PW	65,600	1024
MO	256	41,6

PW: pomegranate wastewater, MO: methyl orange.

Table 4.
COD removal efficiencies of pomegranate wastewater sample and MO solution [30].

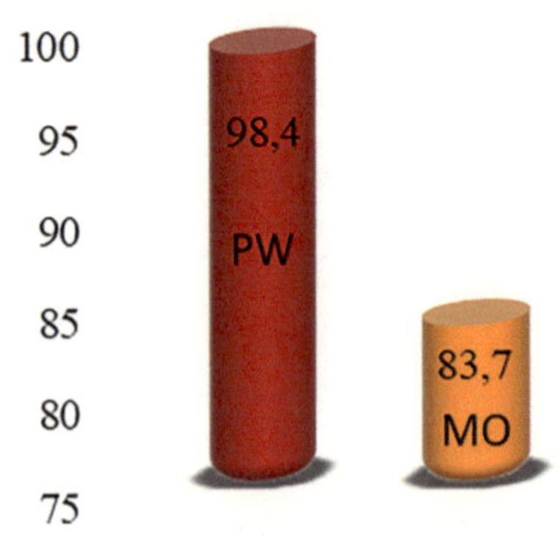

Figure 10.
COD % removal efficiency comparison [30].

Active ingredient	EC_{50}	TU	Classification
Root Length	440,084	0,227	Slightly Toxic
Hypocotyl Length	538,131	0,186	Slightly Toxic

Table 5.
Calculated EC_{50} and toxic unit value for methyl orange initial solution [30].

Active ingredient	EC_{50}	TU	Classification
Root length	450,802	0,222	Slightly toxic
Hypocotyl length	552,409	0,181	Slightly toxic

Table 6.
EC_{50} and toxic unit value calculated for methyl orange solution after plasma discharge [30].

toxicity test result for the real wastewater sample. In this study, although a synthetic solution was used, it should be considered that discharging a real textile wastewater without treatment may increase the toxic effect in wastewater treatment plants.

3.6 Effect of H_2O_2 (hydrogen peroxide) formation

Reference [35] stated in their study that H_2O_2 is one of the most important oxidants produced in the atmospheric pressure plasma system. Reference [36] mentioned in their study that long-lived reactive species such as hydroxyl radical and hydrogen peroxide are formed by plasma discharge. They mentioned that the resulting reagents play an important role in breaking down the contaminants in the liquid contacted in the plasma application and that these radicals can form at the liquid-gas interface or directly in the liquid. Reference [32] stated in oxygen gas fed plasma studies that hydrogen peroxide is an important oxidizing agent (1.78 V oxidation potential), has a longer lifespan than other radicals, and is therefore effective in pollutant removal. They also mentioned that in the aqueous phase, the transferred reactive species interact with the surrounding molecules for a long time; therefore, new reactive species are produced according to their properties, which enable them to participate in chemical reactions. In literature studies, it is mentioned that the most important of the reactive species formed by plasma contact is that hydrogen peroxide has a long life and that oxidation continues for a while after the discharge is stopped in plasma application, thus treatment continues. In this study, the amounts of hydrogen peroxide formed after plasma for MO and pomegranate wastewater sample and pure water for the highest removal efficiencies are given in **Table 7**.

Samples	H_2O_2 amount (mg/L)
MO	25
PW	25
DW	25

PW: pomegranate wastewater, MO: methyl orange, DW: deionized water.

Table 7.
Effect of plasma on H_2O_2 formation [30].

4. Discussion

Within the scope of the study, the atmospheric pressure plasma system, an innovative technology, was preferred for target pollutants. To investigate the color removal efficiency of atmospheric pressure plasma, two different samples, a dye (methyl orange) and a colored wastewater sample (pomegranate wastewater), were analyzed. Oxygen gas was used as feed gas in the study.

Methyl orange color removal study was carried out in three repetitions under optimum conditions (20 mg/L concentration, 5 minutes plasma contact time, 5 minutes mixer time, and 1.5 cm distance and 12 mg/L flow rate). The best color removal was achieved as 98% (other repeat studies were 97%).

$FeSO_4$ was used for catalyst addition. Catalyst studies were carried out under optimum conditions for methyl orange dyestuff. Accordingly, 91% color removal efficiency was achieved after plasma contact with the addition of catalyst. As a result of the study, it was concluded that plasma achieved color removal with high efficiency even without the addition of catalyst.

pH and conductivity measurements were made before and after plasma contact. For all measurements made, it was observed that the pH values after plasma decreased compared to the initial pH. According to the literature research, it can be said that it is important to monitor the pH change of the solution after plasma, as the density of the oxidant produced varies depending on the pH values and can therefore indicate the type of reactions going on during the experiments. For all measurements made in the study, post-plasma conductivity values reached higher values than the initial conductivity values. According to literature research, we can say that plasma discharge increases inorganic ions in water, and as a result, plasma increases the conductivity in water.

In this study, pomegranate wastewater sample was used for colored wastewater experiment. Pomegranate wastewater was prepared at 100%, 25%, 10%, and 5% dilutions, and absorbance was measured before and after plasma. The absorbance value of the raw sample (100%) pre-plasma was 660 nm. While the value of this was read 0.395, it was read 0.916 abs after plasma. Based on these data, it was concluded that plasma is also effective in removing the color of colored fruit wastewater.

The effect of plasma on COD removal for methyl orange solution and pomegranate wastewater was examined. COD removal efficiencies of 88.6% and 83.7% were obtained, respectively. Considering the efficiency, it was seen that plasma was very effective for COD removal, especially in fruit wastewater.

Literature studies indicate that hydrogen peroxide has a longer lifespan than other radicals formed by plasma discharge and plays an important role in oxidizing pollutants. Based on this, hydrogen peroxide formation was examined with kits in the study. In experiments conducted with plasma contact, the highest value on the scale, that is, 25 mg/L hydrogen peroxide formation, was observed for the best removal efficiency.

In addition to color purification using plasma technology, the toxicity of possible by-products was also determined in the study. Toxicity testing was performed for methyl orange. According to the test results, root and hypocotyl lengths were calculated as 0.227 and 0.186, respectively, for the methyl orange starting solution, while they were calculated as 0.222 and 0.181, respectively, for the post-plasma solution. According to the toxic unit classification, all results were evaluated as “Slightly Toxic”.

Considering the results obtained, it was seen that plasma was successful in removing color and COD in liquids with its high reaction speed and energy efficiency.

5. Conclusion

In conclusion, effect of some chemical and physical factors on the process was studied. Response surface methodology was used to optimize the process. Based on the analytical techniques, a degradation pathway was proposed.

Today, the fact that plasmas produced at atmospheric pressure give effective results in applications and are easier and more economical to install than vacuum plasma systems, rapidly increases the interest in these systems. For this reason, by establishing an atmospheric plasma system, a system that can be easily implemented, allowing for minimum chemical use, contributing to green chemistry studies, and not creating waste sludge problems has been established.

This study will contribute to the applications in the field of green chemistry, which have been increasing day by day in recent years, in terms of reducing the chemicals used in advanced oxidation (AOP) processes and thus preventing sludge formation. For this reason in this study, plasma treatment was used as green technique in the water treatment. It is aimed to contribute to the literature with the future applicability of the plasma technology we have created on a laboratory scale in areas related to environmental engineering (cleaning of membrane filters, adsorption and desorption, and so on).

Acknowledgements

This study was supported by Konya Technical University Scientific Research Projects Coordination Office as thesis project number 221001047. The authors are grateful for the financial support provided by Konya Technical University Scientific Research Projects Coordination Office.

Author details

Süheyla Tongur*, Şule Büşra Zambak and Sevil Yıldız
Department of Environmental Engineering, Konya Technical University, Konya, Turkey

*Address all correspondence to: e248128001002@ktun.edu.tr

References

[1] Kumar A et al. Cold atmospheric plasma technology for removal of organic micropollutants from wastewater—A review. The European Physical Journal D. 2021;**75**(11):283

[2] Collivignarelli MC et al. Treatments for color removal from wastewater: State of the art. Journal of Environmental Management. 2019;**236**:727-745

[3] Basha CA et al. Photoelectrocatalytic oxidation of textile dye effluent: Modeling using response surface methodology. Industrial & Engineering Chemistry Research. 2012;**51**(7):2846-2854

[4] Martínez-Huitle CA, Brillas E. Decontamination of wastewaters containing synthetic organic dyes by electrochemical methods: A general review. Applied Catalysis B: Environmental. 2009;**87**(3):105-145

[5] Erkuş A et al. Boya Endüstrisi Atıksularının Karakterizasyonu. Yüzüncü Yıl Üniversitesi Fen Bilimleri Enstitüsü Dergisi. 2018;**23**(3):308-319

[6] Fahmy A et al. Modeling and optimizing acid orange 142 degradation in aqueous solution by non-thermal plasma. Chemosphere. 2018;**210**:102-109

[7] Can OT. COD removal from fruit-juice production wastewater by electrooxidation electrocoagulation and electro-Fenton processes. Desalination and Water Treatment. 2014;**52**(1):65-73

[8] Huang F et al. Degradation of methyl orange by atmospheric DBD plasma: Analysis of the degradation effects and degradation path. Journal of Electrostatics. 2012;**70**(1):43-47

[9] Ong YK et al. Nanofiltration hollow fiber membranes for textile wastewater treatment: Lab-scale and pilot-scale studies. Chemical Engineering Science. 2014;**114**:51-57

[10] Sojka-Ledakowicz J et al. Application of membrane processes in closing of water cycle in a textile dye-house. Desalination. 2010;**250**(2):634-638

[11] Liu Y et al. Carbamazepine removal from water by dielectric barrier discharge: Comparison of ex situ and in situ discharge on water. Chemical Engineering and Processing: Process Intensification. 2012;**56**:10-18

[12] Patil BN, Naik DB, Shrivastava VS. Photocatalytic degradation of hazardous Ponceau-S dye from industrial wastewater using nanosized niobium pentoxide with carbon. Desalination. 2011;**269**(1):276-283

[13] Işık M, Sponza DT. Biological treatment of acid dyeing wastewater using a sequential anaerobic/aerobic reactor system. Enzyme and Microbial Technology. 2006;**38**(7):887-892

[14] Ou W et al. Experimental study on coupling photocatalytic oxidation process and membrane separation for the reuse of dye wastewater. Journal of Water Process Engineering. 2015;**6**:120-128

[15] Qi L, Wang X, Xu Q. Coupling of biological methods with membrane filtration using ozone as pre-treatment for water reuse. Desalination. 2011;**270**(1):264-268

[16] Shin C-H, Bae J-S, Rudolph V. Co-treatment systems combined with unit processes for dye wastewater recycling. Journal of Industrial and Engineering Chemistry. 2014;**20**(2):710-716

[17] Khan TA, Dahiya S, Ali I. Use of kaolinite as adsorbent: Equilibrium, dynamics and thermodynamic studies on the adsorption of Rhodamine B from aqueous solution. Applied Clay Science. 2012;**69**:58-66

[18] Patel D et al. An overview of recent advances in treatment of complex dye-containing wastewater and its techno-economic assessment. Journal of Environmental Management. 2024;**370**:122804

[19] Tao X et al. A novel two-level dielectric barrier discharge reactor for methyl orange degradation. Journal of Environmental Management. 2016;**184**:480-486

[20] Bogaerts A et al. Gas discharge plasmas and their applications. Spectrochimica Acta Part B: Atomic Spectroscopy. 2002;**57**(4):609-658

[21] Fahmy A et al. Plasma polymerized allyl alcohol/O_2 thin films embedded with silver nanoparticles. Thin Solid Films. 2016;**616**:339-347

[22] Vandenbroucke AM et al. Non-thermal plasmas for non-catalytic and catalytic VOC abatement. Journal of Hazardous Materials. 2011;**195**:30-54

[23] Fridman A. Elementary Plasma-Chemical Reactions in Plasma Chemistry. Cambridge University Press; 2008. pp. 12-91

[24] Go DB et al. Ionic winds for locally enhanced cooling. Journal of Applied Physics. 2007;**102**(5):053302

[25] Park GY et al. Atmospheric-pressure plasma sources for biomedical applications. Plasma Sources Science and Technology. 2012;**21**(4):043001

[26] Pârvulescu VI, Magureanu M, Lukes P. Plasma Chemistry and Catalysis in Gases and Liquids. John Wiley & Sons; 2012

[27] Mizuno A. Generation of non-thermal plasma combined with catalysts and their application in environmental technology. Catalysis Today. 2013;**211**:2-8

[28] Samukawa S et al. The 2012 plasma roadmap. Journal of Physics D: Applied Physics. 2012;**45**(25):253001

[29] Locke B, Shih K-Y. Review of the methods to form hydrogen peroxide in electrical discharge plasma with liquid water. Plasma Sources Science and Technology. 2011;**20**:034006

[30] Zambak ŞB. Investigation of the Usage of Plasma Systems in the Field of Environmental Engineering [Thesis]. Turkey: Konya Technical University, Graduate Education Institute; 2023. p. 106

[31] Jiang B et al. Degradation of azo dye using non-thermal plasma advanced oxidation process in a circulatory airtight reactor system. Chemical Engineering Journal. 2012;**204-206**:32-39

[32] Yamada M et al. Nonthermal atmospheric pressure plasma for methylene blue dye decolorization by using slug flow reactor system. Plasma Chemistry and Plasma Processing. 2020;**40**(4):985-1000

[33] Bursać Kovačević D et al. Effects of cold atmospheric gas phase plasma on anthocyanins and color in pomegranate juice. Food Chemistry. 2016;**190**:317-323

[34] Tongur S, Atmaca H. A battery of simple bioassays for domestic and industrial wastewater treatment plants in Konya, Turkey. Sustainability. 2024;**16**(1):316

[35] Chandana L, Lakshminarayana B, Subrahmanyam C. Influence of hydrogen

peroxide on the simultaneous removal of Cr(VI) and methylene blue from aqueous medium under atmospheric pressure plasma jet. Journal of Environmental Chemical Engineering. 2015;**3**(4, Part A):2760-2767

[36] Sanito R, You S-J, Wang Y-F. Degradation of contaminants in plasma technology: An overview. Journal of Hazardous Materials. 2021;**424**:127390

Chapter 4

Perspective Chapter: Role of Nanobubbles Technology in Wastewater Treatments

Angeles Blanco, Guillermo Nieto, Carlos Negro, M. Concepción Monte, Ana Balea, Elena Fuente and Jose Luis Sanchez-Salvador

Abstract

Bulk nanobubble technologies, developed over the last decade, represent an environmentally friendly innovation in water treatment. Nanobubbles (< 200 nm) exhibit unique properties due to their colloidal behavior and negative surface charge. They are affected by Brownian motion, which enables them to remain stable in aqueous systems for several days, leading to a new approach for liquid-gas and solid-gas contact systems. Their main characteristics include exceptional stability, long residence times, large specific surface area, high gas transfer efficiency, strong interfacial potential, enhanced adhesion, and the ability to generate free radicals. These features make nanobubbles highly promising for enhancing conventional water and wastewater treatments, improving the overall efficiency of wastewater treatment plants. Main applications include flotation processes, aerobic biological technologies such as activated sludge, biological activated carbon, or membrane bioreactors, and ozone-based oxidation treatments. The generation of active free radicals that oxidize organic compounds facilitates pollutant degradation. Nanobubbles improve the efficiency of individual treatment processes, reduce operating costs and chemical consumption, and, in summary, may extend the limits of existing wastewater treatment plants. This chapter explores the fundamental properties and technological applications of nanobubbles in wastewater treatment, highlighting their potential to remove macro- and micro pollutants while improving overall process sustainability.

Keywords: nanobubbles, nanotechnology, wastewater treatment plant, nano-ozonation, nano-flotation, nano-aeration, emerging contaminants

1. Introduction

Ensuring safe and high-quality water is a global priority, as freshwater scarcity and pollution continue to intensify due to rapid urbanization and industrial expansion. Traditional water treatment technologies, although effective at removing macropollutants, struggle to fully address the challenges posed by micropollutants such as emerging contaminants (ECs). These pollutants, including pharmaceuticals, pesticides, chlorinated hydrocarbons, and persistent organic matter, among others, are often resistant to conventional physical, chemical, and biological treatments, remaining in the final effluents of wastewater treatment plants (WWTPs), leading to risks of secondary pollution [1, 2]. **Figure 1** summarizes the typical performance of conventional treatments and indicates the extent to which each process can remove ECs.

To extend the efficiency of urban WWTPs, the EU has updated the urban wastewater treatment legislation with the new Directive EU 2024/3019. It mandates quaternary treatments for removing micropollutants in plants serving more than 150,000 population equivalent. This requires the integration of advanced, ad-hoc technologies into existing treatment trains to enhance and complement conventional equipment.

In this context, nanobubble (NB) technology has recently emerged as a promising upgrade for a wide variety of treatments. NBs present unique physicochemical properties, particularly at gas-liquid interfaces, which enable improvements in processes where contact efficiency and mass transfer are critical. The next section explores these properties in detail, highlighting how NBs can enhance current technologies such as flotation, biological treatments, or advanced oxidation, among others.

On the other hand, NB production depends on scalable, energy-efficient technologies such as hydrodynamic cavitation (using devices like Venturi injectors), fluidic oscillation, microporous materials, and ultrasonic irradiation [3–6]. At the industrial scale, production focuses on optimizing generator design, operational parameters, and process control to ensure consistent NB size, stability, and concentration, broadening their applicability. Moreover, recent advances highlight the use of microporous materials to passively produce NBs with tunable gas composition, reducing energy costs compared to traditional mechanical aeration or ultrasound [7].

This chapter aims to explore the fundamental properties and technological applications of nanobubbles (NBs) in wastewater treatment, highlighting their potential to enhance pollutant removal efficiency, energy performance, and process sustainability. The methodology follows the steps described in **Figure 2**. First, a precise search on Web of Science (WoS) was conducted by selecting two main topics: "nanobubble" and "water treatment," including their potential related synonyms like "bulk-nanobubbles" or "wastewater treatment," among others.

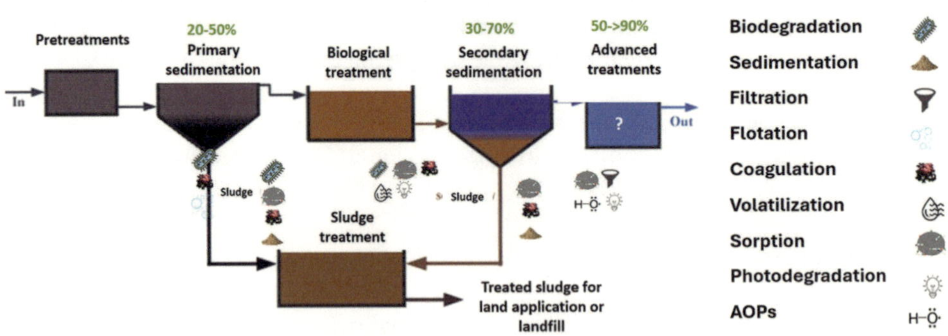

Figure 1.
Potential removal of ECs in a conventional WWTP.

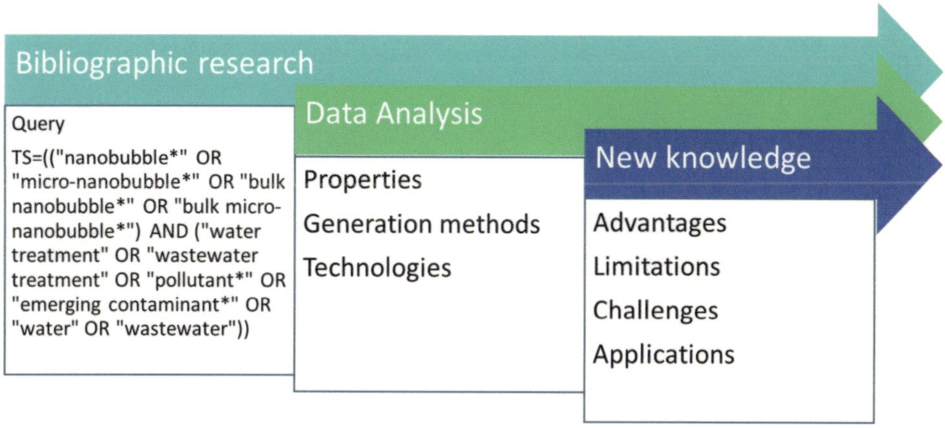

Figure 2.
Methodology for the bibliographic search.

Once the bibliometric parameters were selected, the obtained data were analyzed and categorized into three main topics: nanobubbles' physicochemical properties, generation methods, and effects on water technologies. Finally, a critical analysis of the papers provided key information about nanobubble properties and applications, considering advantages, current limitations, and further challenges.

2. Properties, advantages, and limitations of nanobubbles

NBs, typically defined as gas-filled cavities smaller than 1 μm, exhibit extraordinarily long lifetimes in water, high surface charge (zeta potential), large specific surface area, and enhanced gas transfer efficiency compared to microbubbles (MBs) or conventional aeration systems [8, 9]. Unlike ordinary bubbles, which rapidly rise and burst at the liquid surface, NBs remain stable for days or even weeks. This stability prolongs their interaction with dissolved and colloidal pollutants, as well as microbial communities [10, 11]. NBs are generally classified into two types: surface NBs and bulk NBs [12]. Surface NBs remain attached to liquid–solid interfaces, whereas bulk NBs are spherical nanobubbles dispersed within the liquid phase [13]. In water treatment applications, bulk NBs are the predominant type [12]. **Figure 3** shows a summary of the unique properties of bulk NBs compared to MBs.

As bubble size becomes smaller, the rising velocity decreases (up to 10^6 times lower compared to MBs), and surface tension increases. The outer layer of NBs tends to be negatively charged, depending on the pH and the surrounding chemical environment. This charge is quantified through zeta potential, which can reach - 50 mV under neutral or alkaline pH. As a result, repulsive forces between NBs prevent coalescence and confer remarkable stability and negligible buoyancy. Consequently, NBs remain suspended in water for extended periods (days) without rising quickly to the surface, contributing to their superior performance in pollutant removal and gas-transfer efficiency [14, 15].

At the same time, decreasing the bubble size increases the interfacial surface area, one of the most significant advantages of NBs. The enlarged surface enhances

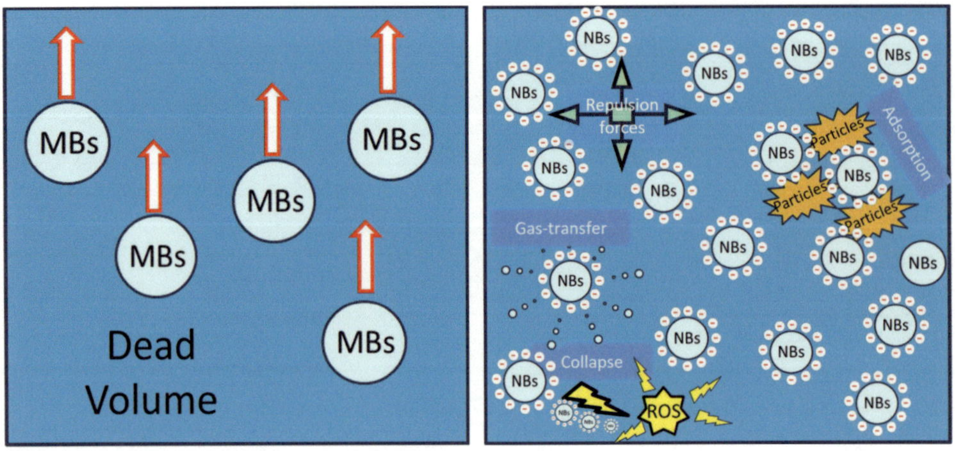

Figure 3.
Properties of microbubbles (MBs) compared to nanobubbles (NBs).

adsorption on suspended and colloidal contaminants. Combined with electrostatic forces, it accelerates the aggregation of oppositely charged particles in the water matrix. This improvement in adsorption kinetics increases removal efficiency and can be effectively combined with biological treatment processes. The convergence of adsorption, biodegradation, and NB-enhanced processes offers a powerful and environmentally friendly strategy for upgrading water treatment. Furthermore, NBs can improve flotation processes by reducing the chemical oxygen demand, turbidity, total solids, and cationic demand of the wastewater [15, 16].

The expanded surface area also facilitates more efficient mass transfer compared to traditional systems, such as aerators in WWTPs [17, 18]. Studies have shown that the volumetric mass transfer coefficient (k_La) and gas utilization efficiency can be more than doubled when NBs are used instead of larger bubbles [15, 17, 18]. These improvements translate into doubling the dissolved oxygen in biological reactors with the same supply, leading to improvements in process efficiency.

In case of collapse, NBs can generate reactive oxygen species (ROS), such as hydroxyl radicals (·OH) or superoxide anions ($O_2{\cdot}^-$), depending on several factors like the chemical species present in the water matrix or the pH, among others. This phenomenon is attributed to the extreme temperature and pressure reached in localized hot spots during bubble implosion (up to 5 GPa, depending on the gas type). Under such conditions, the high concentration of ions accumulated in the disappearing interface does not have enough time to dissolve, releasing the chemical energy required to form ROS. The type of ROS formed depends on the chemical species present in the water. When NBs interact with oxidants such as ozone or hydrogen peroxide, additional radicals are produced, enhancing oxidative degradation [19, 20].

By simultaneously improving oxygen transfer, accelerating adsorption kinetics, and facilitating oxidative degradation, NBs offer synergistic effects that align with the growing demand for sustainable technologies capable of addressing complex mixtures of organic pollutants in municipal and industrial wastewater.

Beyond their physicochemical properties, NBs provide several operational and environmental advantages that make them a promising innovation in water treatment.

Advantages	Limitations
High gas transfer efficiency and enhanced oxygen dissolution	High energy consumption during NB generation
Large specific surface area, improving adsorption and mass transfer	Equipment cost and maintenance requirements
Long-term stability in aqueous systems	Difficulties in scaling up while maintaining bubble stability
Ability to generate reactive oxygen species (ROS) for the oxidation of pollutants	Sensitivity to water chemistry (pH, ionic strength, multivalent cations)
Enhanced coagulation/flocculation and particle aggregation	Lack of standardized measurement and monitoring protocols
Enhanced flotation	Determination of optimal micro/nano ratio
Improved oxygen availability for biological treatment processes	Limited understanding of ROS formation and NB–microorganism interactions
Potential for micropollutant removal	Potential disturbance of microbial community balance
Reduced chemical use and improved process efficiency in pilot-scale applications	Lack of long-term, full-scale validation and economic assessment

Table 1.
Advantages and limitations of nanobubble technologies.

However, despite these advantages, several limitations still hinder the large-scale application of NB technologies. The high energy demand and equipment costs associated with NB generation remain significant barriers to full-scale implementation. Maintaining bubble stability and consistent performance under variable water conditions also poses technical challenges. In addition, the lack of standardized characterization methods and cost-effective monitoring tools limits accurate evaluation of NB behavior and performance. At the biological level, potential alterations in microbial community structure and the limited understanding of long-term impacts require further research. **Table 1** presents a summary of the main advantages and limitations of NB technologies.

3. Technical applications

The different properties of NBs justify that they act through several mechanisms for the removal of pollutants by different technologies, depending on how they interact with the particle, colloid, and dissolved pollutant, as shown in **Figure 4**. The following sections provide a detailed description of the specific applications and benefits of NBs in flocculation, flotation, biological treatments, adsorption, and advanced oxidation processes.

3.1 Flocculation processes

Flocculation is involved in the separation stages of many industrial processes. It improves sedimentation, flotation, filtration, and other separation operations. Some of these processes, such as flotation, can be directly improved by using NBs, but the

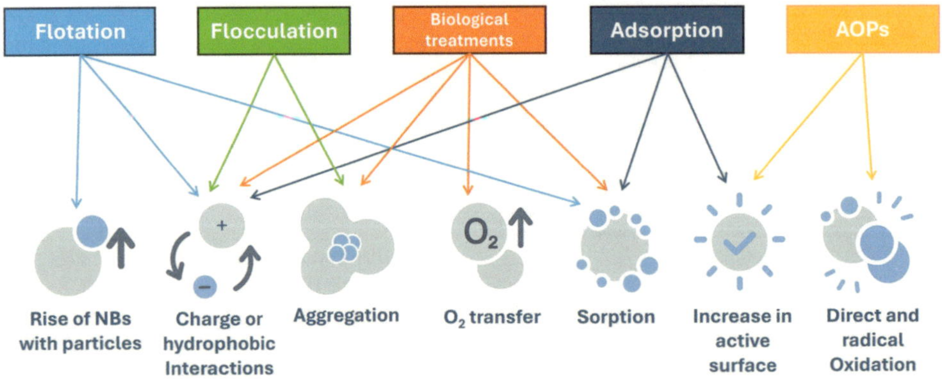

Figure 4.
Main mechanisms of nanobubble interactions in different water treatment technologies.

negative charge of the NBs promotes interactions that also favors flocculation. Therefore, there is a synergistic effect in the separation process due to the use of coagulants/flocculants and the presence of NBs.

Most of the studies on the effect of NBs on particle interactions are in the framework of mineral suspensions and water treatment. Flocculation of coal, kaolin, clay, and calcite, for example, has been improved by NB additions. Most of the particles in these studies were hydrophobic or had hydrophobic parts. For example, kaolin with plate-like morphology has hydrophobic edges, and NBs promote the interaction between edges, leading to a gel-like flocculated suspension [21, 22].

NBs are mainly adsorbed on the hydrophobic particle surfaces [23], although they can also be adsorbed on hydrophilic particles [24, 25]. This adsorption is driven by the interactions described in the DLVO (Derjaguin–Landau–Verwey–Overbeek) theory and by hydrophobic forces [26]. Therefore, electrostatic repulsion and van der Waals attraction forces define the interaction energy profile between the bubble and the solid-liquid interface. These forces determine the system's stability and the likelihood of effective attachment, depending on the physicochemical conditions of the surrounding medium.

The adsorption of NBs onto particle surfaces has mainly four consequences:

- The apparent volume of the particle increases, which promotes its interaction with other particles through orthokinetic flocculation in forced convection systems [27].

- The area of the solid-liquid interface decreases, this reduces the stability of colloidal suspensions [28].

- The NBs covering neighboring particles can interact before the distance between surfaces is small enough to cause significant repulsive forces. This is called long-term forces, and they allow the formation of capillary bridges between particles [21, 29].

- The zeta potential of the suspension is altered in the presence of NBs, which affects stability. This fact has been observed in kaolin suspensions, where NBs neutralized the zeta potential and reduced the stability of fine particles [24].

In summary, the presence of NBs in a suspension may induce the flocculation of particles, improving the efficiency of the process. The main mechanism explaining the improvement is the enhancement of long-term forces or hydrophobic forces due to the hydrophobic interactions, forming capillary bridges between adsorbed NBs on the particle surface [24, 28, 29]. The stability of capillary bridges is related to the charge density of the NB. High charge density decreases bridge stability, which affects floc stability [29]. Since the addition of cationic surfactants and flocculants affects the charge density of NBs, some studies have been carried out with this kind of additive. Wang et al. proved a synergic effect between sodium oleate and NBs in calcite flocculation because of both the decrease in zeta potential and NB size, and the stability increase [30]. Later, Dutta et al. proved and explained this fact while working with different surfactants and salts [29]. Some authors have studied the effect of NBs on the flocculation of kaolin [24, 27] and fine coal [28] induced by polyacrylamide. The studies showed a clear improvement in flocculant efficiency due to the NBs, and in the case of fine coal, some aggregation was driven by NBs before adding the flocculant. These results evidence the possibility of using NBs to improve flocculation efficiency, especially in the case of mineral particles.

3.2 Flotation processes

Flotation is a well-known separation process that uses gas bubbles as carriers to transport suspended particles to the liquid surface [31, 32]. Considering the NBs' properties described above, flotation performance can be significantly enhanced through the following mechanisms:

- Surface modification: Nanobubbles (NBs) adsorb onto particle surfaces, altering surface energy and increasing hydrophobicity. This facilitates stronger particle-bubble interactions and improves separation efficiency [33].

- Heterocoagulation: NBs present a high negative zeta potential, which favors electrostatic interactions with positively charged particles. This leads to heterocoagulation and destabilization of colloidal matter, promoting particle aggregation and separation [6].

- Nucleation sites: NBs serve as nucleation centers for dissolved gases. In the case of a pressure drop or agitation, they expand, forming MBs and improving the probability of particle–bubble collisions [34].

- Bubble–particle attachment efficiency: Due to their nanoscale size, NBs can spread across particle surfaces, reduce induction time, and increase attachment efficiency. During particle–bubble interactions, the presence of surface NBs on particles leads to wetting film rupture at larger thicknesses compared to systems lacking surface NBs. This effect is particularly beneficial for removing micro-sized particles that would be hard to separate using conventional flotation [32, 35].

Flotation with NBs has been successfully applied to more efficiently remove contaminants such as organic matter, metal ions, oils, dyes, minerals, and fine powders [15, 35]. Efficiency depends strongly on bubble size, surface charge, and the contaminant properties. Smaller bubbles lead to higher collection efficiency and improved separation. Han et al. demonstrated that separation efficiency is proportional to the product of the collection efficiency and the number of bubbles present [36].

Several innovations highlight these benefits. Terasaka developed a spiral-flow bubble generator that successfully removed fine carbon particles [37]. Nazari et al. demonstrated that NBs increased the flotation recovery of coarse particles (>100 μm) by 14% (from 53% to 67%) through upgraded bubble–particle attachment and a reduced Reynolds number [38]. Similarly, Calgaroto et al. showed that combining NBs (200–720 nm) with large bubbles (400–800 μm) enhanced the capture of fine and ultrafine quartz particles (<8–74 μm) due to their slower rising velocity and entrainment effect [39].

High removal efficiency was also achieved in oily wastewater [40], where column flotation using NBs with fine bubbles reduced oil content to less than 5 mg/L [16]. Other advances include a T-tube flotation device that decreased oil concentrations on offshore platforms from 38 to 12 g/m^3, while spiral-flow bubble generators recovered 90% of iron oxide fines within 60 minutes [18].

3.2.1 Dissolved air flotation

Dissolved air flotation (DAF) process is the most established fine-bubble-based technology. It produces MBs (20–100 μm) by depressurizing air-saturated water, improving contaminant removal compared to conventional systems [41]. NBs can further enhance DAF through several mechanisms:

- Seeding effect: NBs act as seeds that expand into MBs during decompression, producing a stable and uniform bubble size distribution [6].

- Improved bubble–particle collision: The increased bubble number concentration provided by NBs raises the collision probability between bubbles and particles, including those in the submicron range [34].

- Electrostatic interactions: The negative surface charge of NBs promotes attachment to cationic contaminants, thereby increasing bubble–particle interactions [33].

- Localized ROS generation: In some cases, the collapse of oxygen or ozone NBs generates radicals, such as •OH, which degrade organic contaminants. Hence, hydrophobicity and flotation efficiency are enhanced [42].

Some DAF applications include the removal of oil, metal ions, dyes, and fine particulates. There are studies that report higher efficiency, reduced chemical demand, and shorter flocculation times [41]. For instance, Calgaroto et al. achieved 80% removal of amine pollutants using DAF with NBs [39], while Etchepare et al. achieved > 99% removal of Fe^{3+} and 66–91% removal of $Fe(OH)_3$ colloids [6]. Similarly, Xiao et al. further demonstrated that the presence of NBs inhibits precipitation and increases flotation efficiency, from < 20% to 90% when treating styryl phosphoric acid-Pb particles to recover organic phosphine [33]. Other authors report > 99% oil removal in oil–water separation systems [34]. NB-based flotation is also particularly effective in removing algal cells and emulsified hydrocarbons that are otherwise difficult to treat using conventional DAF [33].

Mixtures of micro and nanobubbles (MNBs) exhibit improved bubble persistence, increasing the reliability and robustness of DAF processes under variable matrix conditions [42]. Hence, clarification is improved, leading to reduced dosages of coagulants and flocculants [32]. Overall, the evidence highlights the powerful role of NBs in enhancing flocculation, pollutant attachment, and mass transfer rates in flotation and DAF processes, turning them into a promising technology for water and wastewater treatment.

3.3 Biological treatments

3.3.1 Aeration

The efficiency of biological wastewater treatment is deeply influenced by the availability of dissolved oxygen to sustain microbial metabolism. Conventional aeration using microbubbles (MBs) is often limited by rapid bubble rise, low oxygen solubility, and high energy consumption. Nanobubbles (NBs) may overcome these limitations due to their specific properties (small size, high surface-to-volume ratio, negligible buoyancy, etc.), allowing them to remain in the water matrix for longer periods. Therefore, oxygen transfer efficiency can nearly double when NBs are used as the aeration medium [43, 44]. It has been reported that dissolved oxygen concentrations increase from 2 to 4.5 mg/L, offering more effective oxygen enrichment for microbial communities in biological systems [43].

Unlike coarse or fine bubbles that break once they have reached the water surface, NBs dissolve gradually until gas equilibrium is achieved. Tekile et al. observed that NBs maintained high dissolved oxygen concentrations long after their formation, providing a more uniform oxygen distribution in the water matrix [44]. This prevents oxygen-depleted "dead zones," which are common in conventional activated sludge and sequencing batch reactors.

The long persistence of NBs ensures that oxygen is delivered more efficiently and uniformly, maintaining and stimulating aerobic microbial metabolism across the entire matrix [43]. In addition, this also enhances the breakdown of pollutants. Several studies highlight the benefits of NBs for microbial activity and pollutant removal. In pilot-scale biological reactors, NB aeration improved the degradation of organic matter, measured as chemical oxygen demand (COD), the nitrification of ammonium (NH_4^+-N), and even the removal of total phosphorus [45]. Additionally, different authors have reported that NB aeration also reduces sludge production through enhanced cell lysis and improved microbial utilization of organic matter [43, 45].

Energy consumption in wastewater treatment is mainly caused by aeration, which accounts for up to 50–70% of the total energy demand. Since nanobubbles (NBs) increase oxygen transfer efficiency, the air supply required to achieve the same dissolved oxygen levels is reduced. Pilot studies and field applications report reductions of 30–40% in aeration energy consumption while maintaining or even improving effluent quality [43]. Furthermore, the generation of sludge is also reduced, and considering that sludge management represents approximately 60% of operational costs in biological treatment, this is a significant advantage due to the lower biomass production.

3.3.2 Biological activated carbon processes

Biological activated carbon (BAC) filtration is a hybrid technology in which biological degradation and adsorption act simultaneously to remove contaminants from water. Activated carbon (AC) is a well-known material that has been used to capture a wide range of contaminants, such as heavy metals, soluble dyes, hydrocarbons, or phenols, among others [46]. Its porous structure and large internal surface area make it a highly effective material with high removal yields. However, adsorption is inherently limited by saturation, after which the process stops unless the AC undergoes regeneration. In BAC systems, biofilms develop on the carbon's surface and within its pores through bacteriological growth on the adsorbed organic matter. This phenomenon facilitates continuous regeneration of the AC without interrupting the process or requiring investment in external regeneration systems. Hence, the dual mechanism of biodegradation and adsorption forms the core strength of BAC filters. The integration of bulk NBs into BAC systems enhances, on the one hand, the biodegradation and adsorption processes and, on the other, the regeneration of the AC.

One of the principal challenges in BAC processes is oxygen transfer. Bacterial metabolism within the active carbon pores depends on how much dissolved oxygen is present. Since these sites are difficult to reach by an external gas input, conventional aeration systems, whether mechanical diffusers or surface aerators, are energy-intensive and inefficient in terms of mass transfer. Nevertheless, NBs may overcome this limitation by supplying high oxygen dissolution rates of up to twice the conventional value. In addition, their negligible buoyancy permits them to remain suspended for long periods, resulting in continuous diffusion of oxygen. Furthermore, their small size of less than 1 μm allows them to reach smaller mesopores. Consequently, this not only enables microbial growth within the AC but also accelerates their metabolism, resulting in the removal of more biodegradable contaminants [17, 47, 48].

From an operational perspective, the combination of NBs in BAC processes could also lead to energy and sustainability benefits, since NB aeration can reduce energy consumption by up to 80% [47]. These savings could make the process more economically suitable for large-scale WWTPs. Furthermore, since NBs can be formed from different gases like oxygen, ozone, or even carbon dioxide, their application may be designed to meet specific treatment goals, whether promoting biodegradation, enhancing oxidation, or improving adsorption.

Furthermore, the ability of NBs to collapse and generate ROS introduces an oxidative pathway that complements both adsorption and biodegradation. Hydroxyl radicals are non-selective and can attack a wide range of recalcitrant pollutants, including those that resist microbial attack. Considering that BAC affects biodegradable matter, the addition of NBs could expand its scope to more recalcitrant contaminants. For instance, ozone NBs have been shown to increase BAC yield by transforming hydrophobic fluorophores into more biodegradable products [49]. Similarly, NB-based AOPs can synergistically decrease toxic species' concentration before adsorption, leading to less biofilm stress and an extension of the filter life.

The active role of NBs in BAC systems also connects to broader adsorption applications. In traditional granular activated carbon (GAC) or powdered activated carbon (PAC) processes, saturation is a limiting factor. NBs can accelerate adsorption rates and upgrade desorption efficiency while lowering the cost of adsorbent regeneration. Their high surface energy and mobility allow them to act as carriers of pollutants toward adsorption sites, while maintaining aerobic conditions that avoid biofouling. Furthermore, the presence of NBs can affect the physicochemical properties of adsorbent surfaces and contaminants, increasing hydrophobic interactions and enhancing adsorption efficiency for compounds that otherwise display low affinity.

3.4 Advanced oxidation processes

3.4.1 General perspective and · OH generation

Advanced oxidation processes (AOPs) are treatment technologies that *in situ* generate highly reactive radicals, such as · OH, to degrade recalcitrant contaminants in water. These processes aim to mineralize recalcitrant pollutants, which are resistant to conventional treatments, often resulting in less toxic byproducts that improve their biodegradability.

The generation of · OH in micro and nanobubbles (NBs) depends on physicochemical and operational conditions. Key parameters include gas composition, bubble size, pH, temperature, dissolved oxidants, pressure, and external stimuli such as ultrasound or UV irradiation [50].

Bubble size plays a fundamental role in · OH generation. Small bubbles (<200 nm) exhibit higher internal pressures due to the Young–Laplace effect, which leads to gas supersaturation and bubble-collapse processes capable of generating · OH radicals [14]. The probability of these interactions occurring increases due to the bubble stability and slow rising velocity, which extend their residence time in the water matrix [15]. In contrast, larger bubbles rise quickly and generate fewer radicals due to lower internal pressure and shorter lifetimes.

The composition of the bubbled gas is one of the key parameters that affect · OH formation. Oxygen NBs tend to produce more · OH than air or inert gases because dissolved oxygen directly takes part in radical formation reactions [16]. Ozone NBs are particularly effective, as O_3 decomposition leads to the secondary formation of · OH through chain reactions. Hydrogen nanobubbles, although less common, exhibit unique redox chemistry that may influence ROS dynamics [51].

pH is another determining factor that influences · OH generation, which is favored under acidic conditions because lower pH enhances the redox potential of oxidants and minimizes radical recombination, thereby increasing the oxidative degradation capacity. Conversely, alkaline conditions often lead to the scavenging of · OH by hydroxide ions and to the formation of less reactive species such as hydroperoxide anions (HO_2^-), resulting in reduced oxidation efficiency. Nevertheless, pH also governs the surface properties of NBs. Under alkaline conditions, NBs acquire a stronger negative surface charge, reflected in a more negative zeta potential. This enhanced electrostatic repulsion stabilizes the NBs, prolonging their lifetime, increasing the coexistence of oxidants and pollutants in the aqueous phase, and, on the other hand, influencing their interactions with cationic charged contaminants [14]. Moreover, the decomposition of oxidants such as ozone and hydrogen peroxide into · OH is accelerated at higher pH values [52]. Conversely, in acidic environments, NBs exhibit reduced stability and diminished ROS yields. Nevertheless, these assumptions must be considered carefully since external factors influence them too. For instance, it has been proven that the generation of · OH can be increased by the decomposition of oxygen molecules and protons. Hence, more acidic conditions and/or high oxygen concentrations might also lead to the generation of · OH.

Temperature affects reaction constants and NBs' stability. High temperatures can reduce stability, leading to bubble collapse, but they also enhance reaction kinetics, promoting · OH generation [53]. Dissolved oxygen concentration is crucial because it provides the source for ROS. Hence, high dissolved oxygen concentration in the form of NBs enhances ROS production and vice versa [20].

Cavitation caused by pressure changes promotes · OH generation. Under high pressure, followed by rapid depressurization, NBs collapse, forming localized hot spots with extreme temperature ($\approx 10^3$ K) and pressure ($\approx$5 GPa) during several picoseconds [54]. Saturation pressure during dissolved air flotation or similar processes is directly correlated with · OH concentration [6]. Optimal pressures ensure continuous NB generation without destabilizing the system.

External energy inputs further enhance · OH production. Ultrasound has been reported as a promoter of · OH production. Acoustic cavitation increases bubble collapse intensity, raising both the frequency and magnitude of radical formation [55]. UV irradiation also enhances radical production, especially when combined with oxygen or ozone NBs, by initiating photolysis reactions that yield · OH [56].

Oxidants such as ozone and hydrogen peroxide highly induce · OH production when they are present with the NBs. Ozone NBs react to form · OH in alkaline water, while hydrogen peroxide undergoes homolytic cleavage under UV or catalytic activation [52]. The introduction of catalytic reactions, such as Fenton-like (Fe^{2+}) or photocatalytic pathways (TiO_2), can further accelerate these processes [57].

Water matrix chemistry must also be considered. Certain ionic species severely affect radical generation, acting as catalysts or inhibitors. High levels of carbonate scavenge · OH, decreasing effective radical concentrations [58]. Nevertheless, transition metals may catalyze secondary radical reactions, increasing · OH formation. Thus, water chemistry must be carefully considered in practical applications. Overall, a good characterization of the water matrix that is being studied, combined with a thorough understanding of the interaction of these parameters, is key to optimizing the application of NBs.

3.4.2 Ozone-based processes

In conventional ozonation, O_3 suffers from a high self-decomposition rate during the transition from the gas to the liquid phase. This leads to low solubility ($\leq$10 mg/L) in water under neutral pH and room temperature, requiring higher O_3 dosages. Moreover, its short half-life and limited mass transfer coefficient further restrict its efficiency. NBs offer a promising alternative by overcoming these limitations. Their high stability, prolonged residence time, and large interfacial area significantly enhance O_3 solubility and mass transfer in aqueous solutions, thereby extending the oxidative activity of O_3 and its derived ROS. As a result, high concentrations of NBs generate ROS-rich microdomains, where pollutants are more likely to encounter oxidative species, accelerating degradation rates of refractory organic pollutants such as pharmaceuticals and surfactants [42, 59].

The enhanced performance of NB-assisted ozonation can be attributed to several mechanisms. First, the large interfacial area of NBs improves the dissolution of oxidizing gases, which is especially beneficial in ozonation and peroxone (O_3/H_2O_2) systems, where higher solubility directly translates into increased $\cdot OH$ generation [59]. Second, the strong negative zeta potential of NBs facilitates electron transfer reactions that accelerate the decomposition of ozone and hydrogen peroxide into ROS [59].

Experimental studies confirm the effectiveness of NB ozonation, reporting more than 80–90% removal of dyes (e.g., methylene blue, reactive black 5) and pharmaceuticals (e.g., tetracycline, carbamazepine). Compared with conventional ozonation, NB-based systems achieve faster degradation kinetics and higher mineralization (TOC removal) [59]. As a result, lower dosages of ozone, hydrogen peroxide, or persulfate are needed to achieve equivalent pollutant degradation [42]. Furthermore, coupling NB ozonation with external stimuli such as UV light, TiO_2 photocatalysis, or Fe^{2+}/Fe^{3+} catalysts produces greater ROS yields due to higher gas solubility and prolonged oxidant lifetime [42].

It is also important to highlight that controlled and sustained ROS production enables more mineralization of pollutants and fewer toxic intermediates than bulk ozonation [59]. In addition, ROS generated by collapsing NBs can inactivate viruses, bacteria, and biofilms, providing a dual advantage of pollutant oxidation and microbial disinfection [42]. Despite these advances, the exact mechanisms of O_3 transport and interfacial reactions during NB-assisted oxidation remain insufficiently understood, warranting further research [50].

3.4.3 Other AOPs processes

NBs hold significant potential to enhance a wide range of AOPs beyond the conventional ozone- or peroxide-based systems. Their application in emerging catalytic pathways, such as peracetic acid, sodium percarbonate, photoelectrocatalysis, and electro-Fenton oxidation, remains largely unexplored. Investigating the potential synergistic effects of NBs in these systems, particularly regarding ROS generation, improved gas–liquid mass transfer, and catalyst activation, could open new avenues for more efficient contaminant degradation [50].

3.5 Adsorption processes

As explained in the previous section on BAC, NB-based technology also influences adsorption mechanisms. NBs significantly enhance adsorption processes in water treatment by increasing adsorption rates, capacities, and pollutant-adsorbent interactions. Their effectiveness depends on water chemistry, adsorbent type, and pollutant properties, making them a promising and cost-effective tool for advanced water purification. A summary of the main mechanisms through which NBs upgrade adsorption is provided below:

- Adsorption rate and capacity: the presence of NBs accelerates the adsorption rate and can substantially increase the adsorption capacity of adsorbents such as biochar or AC. For instance, NBs enhanced the adsorption capacities by up to an order of magnitude compared to conventional methods in the removal of heavy metals and dyes when compared with conventional systems [60–62].

- Pollutant-adsorbent interactions: NBs facilitate closer contact between pollutants and adsorbent surfaces, often forming pollutant–NB colloids that are more readily captured by adsorbents. This mechanism is especially effective for hydrophobic or surface-active contaminants such as polyfluoroalkyl substances (PFAS) and boron [62–64].

- pH and ionic strength sensitivity: NB performance is strongly affected by pH and ionic strength. Low pH values and ionic strengths stabilize NBs and enhance adsorption, while the presence of multivalent cations (e.g., Fe^{3+}) can further increase removal efficiency [62, 64].

- Surface charge and electrostatic effects: NBs can modify the surface charge of species present in the matrix, creating electrostatic interactions that favor adsorption, especially for charged pollutants and functionalized adsorbents such as chitosan-modified carbon [60, 62–64].

Luo et al. showed that the addition of NBs enhances the interfacial interactions of carbon-based adsorbents with conventional PFAS, improving adsorption efficiency across a range of pH values and ion concentrations [64]. Accordingly, Liu et al. demonstrated that perfluoroalkyl acids were more effectively adsorbed onto GAC modified with chitosan when NBs were added, thanks to the improvement in mass transfer and reduced diffusion resistance [63].

Experimental comparisons between adsorption with and without NBs highlight their benefits. Kouvalakidou et al. reported higher pollutant removal and faster kinetics in batch adsorption experiments using NBs [60]. Kyzas et al. observed that, while AC adsorption capacity for lead ions remained similar, adsorption kinetics were dramatically accelerated, up to 366% faster, with NBs [65]. This increase may be a result of electrostatic interactions at the NBs' interface, which attract charged molecules and drive their penetration into the pores of the adsorbent. Such improvements prove that NBs can reduce the equilibrium time for adsorption, promoting faster removal rates and improving the adsorption–regeneration cycles in industrial filters. In practical terms, larger volumes of water can be treated before saturation or breakthrough occurs, increasing operational efficiency.

NBs also enhance adsorption by stabilizing suspensions and changing surface properties. For instance, Ma et al. showed that ion adsorption onto NBs contributes to their stability, indirectly affecting colloidal interactions with adsorbents [66]. This stabilizing effect, coupled with improved dispersion, promotes the adsorption of complex contaminants such as nanoplastics, as reported by Arenas et al. and Ighalo et al. [67, 68].

Apart from upgraded kinetics, NBs can affect water chemistry to promote adsorption. Cerrón-Calle et al. showed that CO_2 NBs changed interfacial reactivity, creating favorable conditions for these processes [69]. Accordingly, Takahashi et al. showed that ROS formation from NBs could enhance oxidative pretreatment, indirectly favoring adsorption yield by reacting large organics into more easily adsorbable species [70].

Compared with traditional adsorption, which is often limited by diffusion and pore accessibility, NBs enhance surface contact, promote turbulence at the microscale, and modify surface charges. Collectively, these mechanisms result in higher adsorption capacities, faster kinetics, and improved operational efficiency. These findings suggest that NB integration can overcome conventional limitations, making adsorption processes more efficient, sustainable, and versatile in addressing ECs.

4. Challenges

4.1 High energy demand for generation

Producing NBs requires specialized equipment, and the choice of generation method strongly influences energy consumption, bubble size, and bubble stability. Common approaches include ultrasound or hydrodynamic cavitation, pressurized dissolution, jet dispersion systems, electrochemical methods, and high-speed shearing impellers. However, many of these processes are energy-intensive. For instance, electro-chemical generation suffers from high energy consumption and low efficiency [71, 72]. Similarly, high-speed shearing impeller systems generate large quantities of NBs but require high power input and often produce unstable bubble sizes [17]. In contrast, devices such as Venturi injectors can improve efficiency, but their energy balance depends heavily on generator design, operational conditions, and scale [72, 73].

Since energy consumption is one of the main barriers to the large-scale application of NBs, strategies to reduce electrical demand during their preparation and optimize operating conditions for NB-assisted technologies must be prioritized. Whole-life cost calculations and energy balance analyses are essential to assess the feasibility of different generation technologies. Furthermore, exploring in situ NB formation during AOPs – such as bubbles produced in plasma oxidation or heterogeneous photocatalytic oxidation – offers an opportunity to reduce external energy input while enhancing ROS yields [50].

4.2 Capital and operational costs

NB generators are typically more complex and expensive than traditional aerators. Diffusion plate systems, though relatively simple to operate, require high-cost devices with micropores that are easily clogged, which include extra maintenance costs [71]. Jet dispersion methods avoid vibrations but still present operational complexity at scale [72]. Consequently, traditional aeration is cheaper and more robust compared to NB generators, which involve higher capital investment and the risk of fouling or scaling, thereby increasing lifecycle costs.

4.3 Scaling challenges

Although the benefits of NBs are well-documented at laboratory and pilot scales, achieving uniform and stable generation at full scale remains difficult. It has been proven that mechanical and chemical methods yield bubbles across a wide distribution, ranging from tens of nanometers to tens of micrometers, complicating consistent application [8, 72]. The absence of a scientific consensus on the stable generation and persistence mechanisms of NBs further hinders large-scale deployment [54]. Some of the main challenges regarding scaling include reactor hydrodynamics, mixing efficiency, and uniform bubble distribution [74].

4.4 Potential interference of ROS with microbial communities

Depending on several factors described above, NBs can produce ROS such as $\cdot OH$, singlet oxygen ($^1O_2$), or superoxide (O_2^-) when they collapse. Nevertheless, results may vary across studies. Some reports indicate significant disinfection and contaminant degradation related to ROS, while others detect negligible ROS activity under similar conditions [19, 75]. This uncertainty makes process control difficult. Excessive ROS levels could damage beneficial bacteria or extracellular enzymes, unbalancing the biological treatment ecosystem [76].

NBs can double dissolved oxygen levels and, under certain conditions in the water matrix, generate ROS. This may benefit aerobic processes, but such conditions may also disturb or inhibit strict microbial species developed in anoxic/anaerobic zones of biological reactors [47]. Alterations in microbial ecology, such as biofilm structural changes or shifts in dominant populations, could negatively impact processes like denitrification or biological phosphorus removal [77]. Therefore, the long-term stability of biofilms under continuous NB exposure must be assessed.

Despite the promising findings regarding NBs combined with biological and adsorption processes, the interaction mechanisms between them and organic contaminants remain incompletely understood. While ionic attraction and interfacial effects explain enhanced kinetics in heavy metal adsorption, organic molecules are structurally more complex and may interact differently with bubble surfaces. For instance, aromatic compounds or hydrophobic dyes may benefit more from increased surface hydrophobicity induced by NBs, while polar compounds may rely on electrostatic attractions. Understanding these dynamics is essential to optimize the global process.

4.5 Sensitivity to water chemistry

The stability and performance of NBs are deeply influenced by water chemistry, including pH, ionic strength, and the presence of chemical species such as surfactants and organic matter. Studies have shown that higher ionic strength and natural organic matter can destabilize NBs, accelerating their collapse and reducing their persistence [78]. This implies that complex water matrices, which contain diverse ions and organic contaminants, may not retain NBs long enough to achieve the expected benefits. Furthermore, changes in zeta potential, depending on gas type and water chemistry, directly influence NBs' stability and their interactions with pollutants [9, 79].

Further research is needed to systematically assess the effects of inorganic species and dissolved organic matter on NB properties, pollutant degradation efficiency, and the formation of potential by-products. Considering the full treatment cycle, it is also critical to evaluate risks associated with undesirable reaction products. For example, in NB-assisted ozonation, high bromide concentrations may lead to bromate (BrO_3^-) formation, posing regulatory and health concerns [50]. Developing strategies to predict and mitigate these risks will be essential for the safe and sustainable application of NB technologies in water and wastewater treatment.

4.6 Measurement and monitoring difficulties

Developing standard methods to precisely control the physicochemical properties of NBs is critical for advancing their application in water treatment. Achieving consistent regulation of parameters such as bubble size and concentration will improve the reproducibility of treatment outcomes and enable broader implementation. To this end, optimizing generation methods and establishing international standards are essential steps. A major challenge is that NBs are typically smaller than the optical resolution of conventional plant instrumentation, making their detection and monitoring difficult. Their characterization often requires advanced tools such as nanoparticle tracking analysis, dynamic light scattering, or zeta potential analyzers [9]. However, these instruments are not commonly available in municipal or industrial plants, limiting the operator's ability to confirm NB presence and stability in real time. Consequently, routine process monitoring and optimization are challenging, underscoring the need for cost-effective, accessible, and standardized monitoring technologies.

5. Conclusions and future prospects

NBs offer significant advantages in water and wastewater treatment due to their enhanced gas transfer efficiency, longer interaction with pollutants, and ability to generate ROS under specific conditions. Their effectiveness in improving flocculation, flotation, aeration in biological treatments, AOPs, and adsorption processes has been demonstrated at both laboratory and pilot scales. Nevertheless, several limitations still hinder their large-scale application. High energy consumption, equipment costs, and the complexity of maintaining stable and consistent performance during scale-up remain critical barriers. Moreover, uncertainties regarding ROS generation mechanisms, the influence of water chemistry on bubble stability, and the lack of cost-effective monitoring technologies complicate operational control. In the case of biological systems, potential long-term disturbances of microbial community balance cannot be underestimated.

Overall, this chapter concludes that NB technologies represent a highly promising yet still emerging approach for sustainable water and wastewater treatment. Future research should focus on optimizing generator design, reducing energy demand, and improving real-time monitoring tools to achieve stable, scalable, and economically viable implementation. Advancing the understanding of NB–microorganism and NB–contaminant interactions will be essential to maximize treatment performance and ensure process stability in biological systems. Importantly, the potential of nanobubbles for the direct removal and degradation of micropollutants remains largely unexplored and deserves systematic investigation. Furthermore, integrating NBs into hybrid systems that combine advanced oxidation, adsorption, and biological processes could yield transformative improvements in both micropollutant removal and circular water management. Finally, the establishment of standardized testing protocols, operational guidelines, and life-cycle assessments will be key to ensuring the safe, efficient, and sustainable deployment of NB-based technologies at full scale.

Acknowledgments

The authors are grateful for the financial support from the Spanish Ministry of Science, Innovation and Universities (project PID2022-136816OB-I00, funded by MICIU/AEI/10.13039/501100011033 and by FEDER, UE; TED2021-130147B-C21 funded by MCIN/AEI/10.13039/501100011033 and by NextGenerationEU/PRTR.

Conflict of Interest

The authors declare no conflict of interest.

Author details

Angeles Blanco*, Guillermo Nieto, Carlos Negro, M. Concepción Monte, Ana Balea, Elena Fuente and Jose Luis Sanchez-Salvador
Cellulose, Paper, and Advanced Water Treatment Research Group (CPYTAA), Department of Chemical Engineering and Materials, Complutense University of Madrid, Madrid, Spain

*Address all correspondence to: ablanco@ucm.es

References

[1] Temesgen T, Bui TT, Han M, T-i K, Park H. Micro and nanobubble technologies as a new horizon for water-treatment techniques: A review. Advances in Colloid and Interface Science. 2017;**246**:40–51. DOI: 10.1016/j.cis.2017.06.011

[2] Priyadarshini M, Das I, Ghangrekar MM, Blaney L. Advanced oxidation processes: Performance, advantages, and scale-up of emerging technologies. Journal of Environmental Management. 2022;**316**:115295. DOI: 10.1016/j.jenvman.2022.115295

[3] Yasuda K, Matsushima H, Asakura Y. Generation and reduction of bulk nanobubbles by ultrasonic irradiation. Chemical Engineering Science. 2019;**195**:455–461. DOI: 10.1016/j.ces.2018.09.044

[4] Zimmerman WB, Tesař V, Bandulasena HH. Towards energy efficient nanobubble generation with fluidic oscillation. Current Opinion in Colloid & Interface Science. 2011;**16** (4):350–356. DOI: 10.1016/J.COCIS.2011.01.010

[5] Li T, Cui Z, Sun J, Jiang C, Li G. Generation of bulk nanobubbles by self-developed venturi-type circulation hydrodynamic cavitation device. Langmuir. 2021;**37**(44): 12952–12960. DOI: 10.1021/acs.langmuir.1c02010

[6] Etchepare R, Azevedo A, Calgaroto S, Rubio J. Removal of ferric hydroxide by flotation with micro and nanobubbles. Separation and Purification Technology. 2017; **184**:347–353. DOI: 10.1016/j.seppur.2017.05.014

[7] Ouyang L, Cha H, Zhang J, Hansen HH, Li Q, Tan BH, et al Energy-efficient nanobubble generation with microporous materials. Journal of Molecular Liquids. 2025;**423**:126873

[8] Azevedo A, Etchepare R, Calgaroto S, Rubio J. Aqueous dispersions of nanobubbles: Generation, properties and features. Minerals Engineering. 2016;**94**:29–37. DOI: 10.1016/j.mineng.2016.05.001

[9] Ushikubo FY, Enari M, Furukawa T, Nakagawa R, Makino Y, Kawagoe Y, et al. Zeta-potential of micro-and/or nano-bubbles in water produced by some kinds of gases. IFAC Proceedings Volumes. 2010;**43**(26):283–288. DOI: 10.3182/20101206-3-JP-3009.00050

[10] Seddon JR, Lohse D, Ducker WA, Craig VS. A deliberation on nanobubbles at surfaces and in bulk. ChemPhysChem. 2012;**13**(8):2179–2187. DOI: 10.1002/cphc.201100900

[11] Qiu J, Zou Z, Wang S, Wang X, Wang L, Dong Y, et al. Formation and stability of bulk nanobubbles generated by ethanol–water exchange. ChemPhysChem. 2017;**18**(10):1345–1350. DOI: 10.1002/cphc.201700010

[12] Jia M, Farid MU, Kharraz JA, Kumar NM, Chopra SS, Jang A, et al Nanobubbles in water and wastewater treatment systems: Small bubbles making big difference. Water Research. 2023;**245**:120613

[13] Zhou S, Marcelino KR, Wongkiew S, Sun L, Guo W, Khanal SK, et al. Untapped potential: Applying microbubble and nanobubble technology in water and wastewater

treatment and ecological restoration. ACS ES&T Engineering. 2022;**2**(9):1558–1573. DOI: 10.1021/acsestengg.2c00117

[14] Takahashi M. ζ potential of microbubbles in aqueous solutions: Electrical properties of the gas- water interface. The Journal of Physical Chemistry B. 2005;**109**(46): 21858–21864. DOI: 10.1021/jp0445270

[15] Agarwal A, Ng WJ, Liu Y. Principle and applications of microbubble and nanobubble technology for water treatment. Chemosphere. 2011;**84**(9):1175–1180. DOI: 10.1016/j.chemosphere.2011.05.054

[16] Liu S, Kawagoe Y, Makino Y, Oshita S. Effects of nanobubbles on the physicochemical properties of water: The basis for peculiar properties of water containing nanobubbles. Chemical Engineering Science. 2013;**93**:250–256. DOI: 10.1016/j.ces.2013.02.004

[17] Li H, Hu L, Song D, Al-Tabbaa A. Subsurface transport behavior of micro-nano bubbles and potential applications for groundwater remediation. International Journal of Environmental Research and Public Health. 2014;**11**(1):473–486. DOI: 10.3390/ijerph110100473

[18] Zhou W, Niu J, Xiao W, Ou L. Adsorption of bulk nanobubbles on the chemically surface-modified muscovite minerals. Ultrasonics Sonochemistry. 2019;**51**:31–39. DOI: 10.1016/j.ultsonch.2018.10.021

[19] Takahashi M, Chiba K, Li P. Free-radical generation from collapsing microbubbles in the absence of a dynamic stimulus. The Journal of Physical Chemistry B. 2007;**111**(6):1343–1347. DOI: 10.1021/jp0669254

[20] Ebina K, Shi K, Hirao M, Hashimoto J, Kawato Y, Kaneshiro S, et al. Oxygen and air nanobubble water solution promote the growth of plants, fishes, and mice. PLoS One. 2013;**8**(6): e65339. DOI: 10.1371/journal.pone.0065339

[21] Wang Y, Jiang W, Qin W, Pan Z, Jia L, Hong X, et al The origin of interparticle aggregation: Probing the long-range hydrophobic forces between particles induced by nanobubbles in aqueous solutions. Minerals Engineering. 2024;**218**:109030

[22] Żbik M, Horn RG. Hydrophobic attraction may contribute to aqueous flocculation of clays. Colloids and Surfaces A: Physicochemical and Engineering Aspects. 2003;**222**(1-3):323–328. DOI: 10.1016/S0927-7757(03)00250-4

[23] Knüpfer P, Ditscherlein L, Peuker UA. Nanobubble enhanced agglomeration of hydrophobic powders. Colloids and Surfaces A: Physicochemical and Engineering Aspects. 2017;**530**:117–123. DOI: 10.1016/j.colsurfa.2017.07.056

[24] Li Y, Ma G, Bilal M, Sha J, Bu X. Effect of Bulk Nanobubbles on the Flocculation and Filtration Characteristics of Kaolin Using Cationic Polyacrylamide. Minerals. 2024;**14**(4):405. DOI: 10.3390/min14040405

[25] Li C, Xu M, Zhang H. The interactions between coal particles with different hydrophobicity and bulk nanobubbles in natural water. International Journal of Coal Preparation and Utilization. 2022. DOI: 10.1080/19392699.2019.1618283

[26] Dutta N, Mitra S, Nirmalkar N. Influences of nanobubbles on particle–particle and bubble–particle interactions: A review. Chemical Engineering Research and Design. 2024;**210**:247–270. DOI: 10.1016/j.cherd.2024.08.041

[27] Sha J, Yuan S, Chen X, Chen W, Wang J, Alheshibri M, et al. Effect of bulk nanobubbles on flocculation of kaolin in the presence of cationic polyacrylamide. Physicochemical Problems of Mineral Processing. 2024;**60**(2). DOI: 10.3390/min14040405

[28] Ma G, Cui J, Li J, Bu X, Shao H, Wang Z, et al Nanobubble enhanced flocculation and filtration of fine coal and associated mechanisms. Fuel. 2025;**393**:134989

[29] Dutta N, Mitra S, Nirmalkar N. Understanding the role of surface charge on nanobubble capillary bridging during particle–particle interaction. Langmuir. 2024;**40**(8):4475–4488. DOI: 10.1021/acs.langmuir.3c03963

[30] Wang Y, Wei D, Qin W, Jiao F, Luo X, Pan Z. Effect of nanobubbles on particle flocculation in sodium oleate-calcite flotation system. Minerals Engineering. 2023;**204**:108438. DOI: 10.1016/j.mineng.2023.108438

[31] Kyzas GZ, Matis KA. Flotation in water and wastewater treatment. Processes. 2018;**6**(8):116. DOI: 10.3390/pr6080116

[32] Rubio J, Souza M, Smith R. Overview of flotation as a wastewater treatment technique. Minerals Engineering. 2002;**15**(3):139–155. DOI: 10.1016/S0892-6875(01)00216-3

[33] Xiao W, Ke S, Quan N, Zhou L, Wang J, Zhang L, et al. The role of nanobubbles in the precipitation and recovery of organic-phosphine-containing beneficiation wastewater. Langmuir. 2018;**34**(21):6217–6224. DOI: 10.1021/acs.langmuir.8b01123

[34] Kim M-S, Han M, Kim T-I, Lee J-W, Kwak D-H. Effect of nanobubbles for improvement of water quality in freshwater: Flotation model simulation. Separation and Purification Technology. 2020;**241**:116731. DOI: 10.1016/j.seppur.2020.116731

[35] Li C, Zhang H. A review of bulk nanobubbles and their roles in flotation of fine particles. Powder Technology. 2022;**395**:618–633. DOI: 10.1016/j.powtec.2021.10.004

[36] Han M, Kim W, Dockko S. Collision efficiency factor of bubble and particle (abp) in DAF: Theory and experimental verification. Water Science and Technology. 2001;**43**(8):139–144. DOI: 10.2166/wst.2001.0484

[37] Terasaka K. Separation of fine particles suspended in water using microbubble flotation. Japanese Journal of Multiphase Flow. 2007;**21**(1):77–83. DOI: 10.3811/pmfr.3.43

[38] Nazari S, Shafaei S, Shahbazi B, Chelgani SC. Study relationships between flotation variables and recovery of coarse particles in the absence and presence of nanobubble. Colloids and Surfaces A: Physicochemical and Engineering Aspects. 2018;**559**:284–288. DOI: 10.1016/j.colsurfa.2018.09.066

[39] Calgaroto S, Wilberg K, Rubio J. On the nanobubbles interfacial properties and future applications in flotation. Minerals Engineering.

2014;**60**:33–40. DOI: 10.1016/j.mineng.2014.02.002

[40] Shen W, Mukherjee D, Koirala N, Hu G, Lee K, Zhao M, et al. Microbubble and nanobubble-based gas flotation for oily wastewater treatment: A review. Environmental Reviews. 2022;**30**(3):359–379. DOI: 10.1139/er-2021-0127

[41] Edzwald JK. Dissolved air flotation and me. Water Research. 2010;**44**(7):2077–2106. DOI: 10.1016/j.watres.2009.12.040

[42] Fan W, Zhou Z, Wang W, Huo M, Zhang L, Zhu S, et al Environmentally friendly approach for advanced treatment of municipal secondary effluent by integration of micro-nano bubbles and photocatalysis. Journal of Cleaner Production. 2019;**237**:117828

[43] Ahmadi M, Nabi Bidhendi G, Torabian A, Mehrdadi N. Effects of nanobubble aeration in oxygen transfer efficiency and sludge production in wastewater biological treatment. Journal of Advances in Environmental Health Research. 2018;**6**(4):225–233. DOI: 10.22102/jaehr.2018.116201.1083

[44] Tekile A, Kim I, Lee J-Y. Applications of ozone micro-and nanobubble technologies in water and wastewater treatment. 상하수도학회지. 2017;**31**(6):481–490. DOI: 10.11001/jksww.2017.31.6.481

[45] Leyva M, Flores JV. Reduction of COD and TSS of waste effluents from a sugar industry through the use of air micro-nanobubbles. Journal of Nanotechnology. 2018;**2**(1):7–12. DOI: 10.32829/nanoj.v2i110.32829/nanoj.v2i1.45

[46] Rattier M, Reungoat J, Keller J, Gernjak W. Removal of micropollutants during tertiary wastewater treatment by biofiltration: Role of nitrifiers and removal mechanisms. Water Research. 2014;**54**:89–99. DOI: 10.1016/j.watres.2014.01.030

[47] Xiao W, Xu G. Mass transfer of nanobubble aeration and its effect on biofilm growth: Microbial activity and structural properties. Science of the Total Environment. 2020;**703**:134976. DOI: 10.1016/j.scitotenv.2019.134976

[48] Rameshkumar C, Senthilkumar G, Subalakshmi R, Gogoi R. Generation and characterization of nanobubbles by ionization method for wastewater treatment. Desalination and Water Treatment. 2019;**164**:98–101. DOI: 10.5004/dwt

[49] Wang W-L, Cai Y-Z, Hu H-Y, Chen J, Wang J, Xue G, et al Advanced treatment of bio-treated dyeing and finishing wastewater using ozone-biological activated carbon: A study on the synergistic effects. Chemical Engineering Journal. 2019;**359**:168–175

[50] Ning R, Yu S, Li L, Snyder SA, Li P, Liu Y, et al Micro and nanobubbles-assisted advanced oxidation processes for water decontamination: The importance of interface reactions. Water Research. 2024;**265**:122295

[51] Yang X, Chen L, Oshita S, Fan W, Liu S. Mechanism for enhancing the ozonation process of micro-and nanobubbles: Bubble behavior and interface reaction. ACS ES&T Water. 2023;**3**(12):3835–3847. DOI: 10.1021/acsestwater.3c00031

[52] Staehelin J, Hoigne J. Decomposition of ozone in water: Rate of initiation by hydroxide ions and

hydrogen peroxide. Environmental Science & Technology. 1982;**16**(10):676–681. DOI: 10.1021/es00104a009

[53] Soyluoglu M, Kim D, Zaker Y, Karanfil T. Stability of oxygen nanobubbles under freshwater conditions. Water Research. 2021;**206**:117749. DOI: 10.1016/j.watres.2021.117749

[54] Yasui K, Tuziuti T, Kanematsu W. Extreme conditions in a dissolving air nanobubble. Physical Review E. 2016;**94**(1):013106. DOI: 10.1103/PhysRevE.94.013106

[55] Ashokkumar M. The characterization of acoustic cavitation bubbles–an overview. Ultrasonics Sonochemistry. 2011;**18**(4):864–872. DOI: 10.1016/j.ultsonch.2010.11.016

[56] Gao Y, Duan Y, Fan W, Guo T, Huo M, Yang W, et al. Intensifying ozonation treatment of municipal secondary effluent using a combination of microbubbles and ultraviolet irradiation. Environmental Science and Pollution Research. 2019;**26**(21):21915–21924. DOI: 10.1007/s11356-019-05554-8

[57] Zhang S, Sun M, Hedtke T, Deshmukh A, Zhou X, Weon S, et al. Mechanism of heterogeneous Fenton reaction kinetics enhancement under nanoscale spatial confinement. Environmental Science & Technology. 2020;**54**(17):10868–10875. DOI: 10.1021/acs.est.0c02192

[58] Buxton GV, Greenstock CL, Phillips Helman W, Ross AB. Critical review of rate constants for reacitons of hydrated electrons. Journal of Physical and Chemical Reference Data. United States. 1988;**17**(2): DOI: 10.1063/1.555805

[59] Xia Z, Hu L. Treatment of organics contaminated wastewater by ozone micro-nano-bubbles. Water. 2018;**11**(1):55. DOI: 10.3390/w11010055

[60] Kouvalakidou SL, Varoutoglou A, Alibrahim KA, Alodhayb AN, Mitropoulos AC, Kyzas GZ. Batch adsorption study in liquid phase under agitation, rotation, and nanobubbles: Comparisons in a multi-parametric study. Environmental Science and Pollution Research. 2023;**30**(53):114032–114043. DOI: 10.1007/s11356-023-30342-w

[61] Kyzas GZ, Favvas EP, Kostoglou M, Mitropoulos AC. Effect of agitation on batch adsorption process facilitated by using nanobubbles. Colloids and Surfaces A: Physicochemical and Engineering Aspects. 2020;**607**:125440. DOI: 10.1016/j.colsurfa.2020.125440

[62] Liao L, Chen H, He C, Dodbiba G, Fujita T. Boron removal in aqueous solutions using adsorption with sugarcane bagasse biochar and ammonia nanobubbles. Materials. 2024;**17**(19):4895. DOI: 10.3390/ma17194895

[63] Liu W, Lin T, Zhang X, Jiang F, Yan X, Chen H. Adsorption of perfluoroalkyl acids on granular activated carbon supported chitosan: Role of nanobubbles. Chemosphere. 2022;**309**:136733. DOI: 10.1016/j.chemosphere.2022.136733

[64] Luo Y, Fu K, Wang D, Luo J. Nanobubble-Driven Interfacial Interactions of Carbon-Based Adsorbents with Legacy PFAS: Impact of Concentration, pH, and Coexisting Ions. Environmental Science &

Technology. 2025;**59**(13):6900–6914. DOI: 10.1021/acs.est.4c14726

[65] Kyzas GZ, Bomis G, Kosheleva RI, Efthimiadou EK, Favvas EP, Kostoglou M, et al Nanobubbles effect on heavy metal ions adsorption by activated carbon. Chemical Engineering Journal. 2019;**356**:91–97

[66] Ma X, Li M, Pfeiffer P, Eisener J, Ohl C-D, Sun C. Ion adsorption stabilizes bulk nanobubbles. Journal of Colloid and Interface Science. 2022;**606**:1380–1394. DOI: 10.1016/j.jcis.2021.08.101

[67] Arenas LR, Gentile SR, Zimmermann S, Stoll S. Nanoplastics adsorption and removal efficiency by granular activated carbon used in drinking water treatment process. Science of the Total Environment. 2021;**791**:148175. DOI: 10.1016/j.scitotenv.2021.148175

[68] Ighalo JO, Yap P-S, Iwuozor KO, Aniagor CO, Liu T, Dulta K, et al Adsorption of persistent organic pollutants (POPs) from the aqueous environment by nano-adsorbents: A review. Environmental Research. 2022;**212**:113123

[69] Cerrón-Calle GA, Magdaleno AL, Graf JC, Apul OG, Garcia-Segura S. Elucidating CO2 nanobubble interfacial reactivity and impacts on water chemistry. Journal of Colloid and Interface Science. 2022;**607**:720–728. DOI: 10.1016/j.jcis.2021.09.033

[70] Takahashi M, Shirai Y, Sugawa S. Free-radical generation from bulk nanobubbles in aqueous electrolyte solutions: ESR spin-trap observation of microbubble-treated water. Langmuir. 2021;**37**(16):5005–5011. DOI: 10.1021/acs.langmuir.1c00469

[71] Maeda Y, Hosokawa S, Baba Y, Tomiyama A, Ito Y. Generation mechanism of micro-bubbles in a pressurized dissolution method. Experimental Thermal and Fluid Science. 2015;**60**:201–207. DOI: 10.1016/j.expthermflusci.2014.09.010

[72] Wu M, Song H, Liang X, Huang N, Li X. Generation of micro-nano bubbles by self-developed swirl-type micro-nano bubble generator. Chemical Engineering and Processing-Process Intensification. 2022;**181**:109136. DOI: 10.1016/j.cep.2022.109136

[73] Wu M, Yuan S, Song H, Li X. Micro-nano bubbles production using a swirling-type venturi bubble generator. Chemical Engineering and Processing-Process Intensification. 2022;**170**:108697. DOI: 10.1016/j.cep.2021.108697

[74] Sakr M, Mohamed MM, Maraqa MA, Hamouda MA, Hassan AA, Ali J, et al. A critical review of the recent developments in micro–nano bubbles applications for domestic and industrial wastewater treatment. Alexandria Engineering Journal. 2022;**61**(8):6591–6612. DOI: 10.1016/j.aej.2021.11.041

[75] Mezule L, Tsyfansky S, Yakushevich V, Juhna T. A simple technique for water disinfection with hydrodynamic cavitation: Effect on survival of Escherichia coli. Desalination. 2009;**248**(1-3):152–159. DOI: 10.1016/j.desal.2008.05.051

[76] Wang W, Fan W, Huo M, Zhao H, Lu Y. Hydroxyl radical generation and contaminant removal from water by the collapse of microbubbles under different hydrochemical conditions. Water, Air, & Soil Pollution. 2018;**229**(3):86. DOI: 10.1007/s11270-018-3745-x

[77] Xiao W, Xu G, Li G. Effect of nanobubble application on performance and structural characteristics of microbial aggregates. Science of the Total Environment. 2021;**765**:142725. DOI: 10.1016/j.scitotenv.2020.142725

[78] Meegoda JN, Aluthgun Hewage S, Batagoda JH. Stability of nanobubbles. Environmental Engineering Science. 2018;**35**(11):1216–1227. DOI: 10.1089/ees.2018.0203

[79] Jia W, Ren S, Hu B. Effect of water chemistry on zeta potential of air bubbles. International Journal of Electrochemical Science. 2013;**8**(4):5828–5837. DOI: 10.1016/S1452-3981(23)14725-0

Chapter 5

Turbulent Flow Hydrodynamics CFD Modeling of a Helical Geometry for Pollutant Degradation in Photocatalytic Reactors

Diana Barraza-Jiménez, Derian Manuel Lerma-Mancinas, Sandra Iliana Torres-Herrera, Hugo Iván Flores-Hidalgo, Elva Marcela Coria Quiñones, Sayda Dinorah Coria-Quiñones and Manuel Alberto Flores-Hidalgo

Abstract

A photocatalytic reactor is a device mostly used for degrading the contaminants in aqueous solutions, as well as may be used for contaminant removal from the air, wastewater treatment, and water splitting. Fundamental concepts regarding the design process of such a device are analyzed in this chapter, which includes concepts related with sunlight usage as a clean energy source, which is known as a highly desirable renewable technology. The main interest of this proposal relies on modeling the hydrodynamics of a helical pipeline that will be used in a photocatalytic reactor subject to a turbulent flow regime where sunlight is used to activate a chemical reaction to degrade water pollutants. Numerical calculations are based in computational fluid dynamics (CFD) using ANSYS-Fluent(TM). The different steps, geometric domain, preprocessing steps, setup, and postprocessing steps are described to analyze a numerical calculation for this kind of reactors. The turbulent flow models used are briefly explained so the design intent is better understood by the reader. All the modeling, numerical calculations and analysis are carried over in ANSYS-Fluent(TM), a powerful multi-physics program suite suitable for hydrodynamics simulations.

Keywords: hydrodynamics, ANSYS(TM), photocatalysis, chemical reactor, turbulent flow

1. Introduction

The scarcity of drinking water available for human consumption is one of the biggest concerns in the world. According to the literature, there are around 673 million people in the world who do not have access to clean water, and more than 2.7 billion people face water shortages for at least 1 month a year [1–12]. Water use has increased at twice the rate of global population growth. At the current rate, nearly 2/3 of the world could face water shortages by 2025 [13]. Water treatment is a fundamental part of the solution to water scarcity, and this work joins the efforts to improve the technology options to treat water [14–18]. **Figure 1** shows a diagram illustrating water treatment for domestic use in a general water treatment plant process.

There is a direct relationship between the mitigation of clean water scarcity and the availability and efficiency of water treatment methods. Thus, water treatment science and technology play a key role in solving water shortages in the world. Nowadays, some technologies to treat water are desalination plants [6], chemical methods such as chemical precipitation, coagulation and flocculation, and alternative methods [8], which are among the more studied methodologies. **Figure 1** depicts household water treatment options—filtration, solar disinfection, boiling and chlorination—representative of scalable interventions that enhance drinking-water safety at the population level [14]. Also, photocatalytic reactors fall in the alternative methods that may work well at household level and represent an increased potential due to the possibility of using clean energy sources like solar energy during the water treatment process.

In this context, a design of a photocatalytic reactor for household use is proposed on this work but, due to the complexity required when a reactor is designed, this chapter's scope is limited to the geometry and its hydrodynamics analysis for the proposed reactor. Thus, a helical pipeline is considered the main photocatalytic reactor component because it is where the fluid will be subject to the chemical reactions and light source action. Therefore, the helical pipeline hydrodynamics is the main topic discussed in this work. Computational fluid dynamics (CFD) methodology is applied using the ANSYS-Fluent [19] program to simulate the selected geometry hydrodynamics.

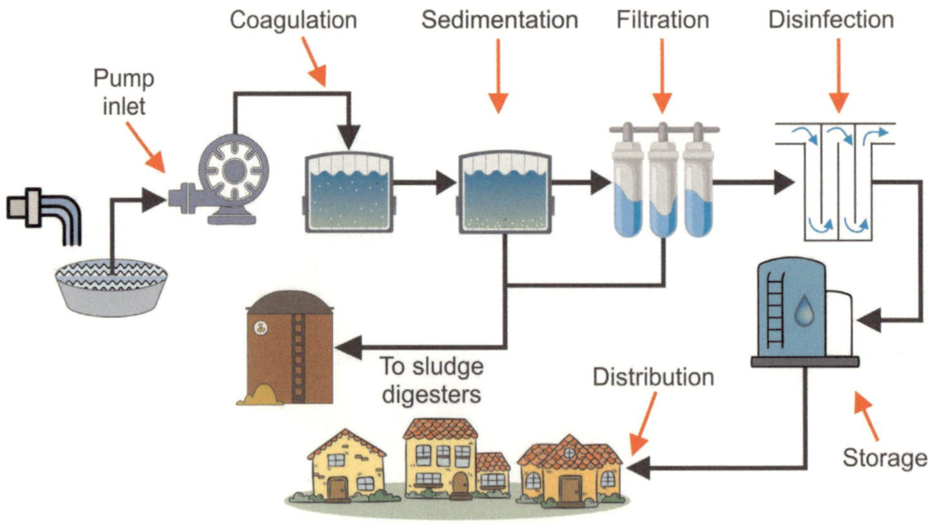

Figure 1.
A water treatment process. Adapted from Ref. [8].

2. Reactors design fundamentals

The design of a chemical reactor involves several steps and disciplines, so it may be considered a complex task that requires multidisciplinary expertise [20–22]. The topic of interest in this section relies in the design of reactors applied in a water purification or treatment process, so we will refer to such specific type of reactors in the next sections. A reactor may be considered a complex device that can be developed in a lot of ways and that can be assembled with multiple parts [20–22]. If we focus on the central part of the reactor, that means the part of the reactor where the chemical reaction occurs one may notice that the parameters that need to be considered reduce to a more manageable amount. Also, the reactor type may be one of the first important design considerations that need to be resolved. From a perspective of a continuous reactor, several authors have reported the flow rate and residence time as the most evaluated operational parameters inherent to a continuous process [23–25]. It is also important to notice that there is no universal design, but instead one needs to consider that reactor designs have been developed for each pollutant.

3. CFD for reactor design

In this case, the reactor design process focuses on hydrodynamics studies, which depend on geometry design and initial considerations of the reactor type. Mostly, the proposal within this document relies on the use of CAD/CAE tools since hydrodynamics simulations are carried over computational fluid dynamics (CFD) software. CFD simulation has become a powerful tool in the design development process since its simulation capabilities have increased in late years. A brief view of the CFD simulation capabilities may be observed in **Figure 2**.

Computational fluid dynamics (CFD) may be thought of as the science of using computers to predict liquid and gas flows based on the governing equations of conservation of mass, momentum, and energy. Through CFD, one may analyze, understand, and predict the fluids that make up nearly every part of our world.

With CFD, the user applies the science laws and rules for predicting fluid flow, heat and mass transfer, chemical reactions, and related phenomena by solving numerically the set of governing mathematical equations:

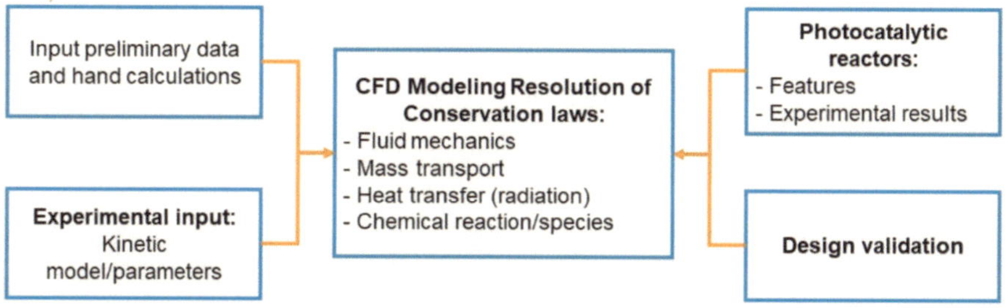

Figure 2.
CFD numerical computation role in reactor design process. Adapted from Ref. [20].

- Conservation of mass.
- Conservation of momentum.
- Conservation of energy.
- Conservation of species.
- Effects of body forces.

4. Brief presentation of the governing equations

The fundamental governing equations in CFD refer to the continuity, momentum, and energy equations as mentioned in the final part of the previous section. For a flow where we have water, the part of the governing equations as displayed in Eqs. (1)–(10) that can be considered for the mass flow [26] through the system are shown next:

$$\frac{\partial}{\partial t}\left(\phi S_w \rho_w\right) = \nabla . \left[\frac{\kappa K_{rw}}{\mu_w}\rho_w\left(\nabla P_w - \rho_w g \nabla D\right)\right] + q_w \tag{1}$$

According to the ANSYS Fluent manual [27], Eq. (2) is the general form of the mass conservation equation and is valid for incompressible as well as compressible flows. The source is the mass added to the continuous phase from the dispersed second phase (for example, due to vaporization of liquid droplets) and any user-defined sources. Eq. (2) is as shown:

$$\frac{\partial p}{\partial t} + \nabla . \rho(\vec{v_r}) = 0 \tag{2}$$

For momentum conservation, the following equations can be considered as the governing equations [26] for this concept:

$$\frac{\partial u}{\partial t} + u\frac{\partial u}{\partial x} + v\frac{\partial u}{\partial y} + w\frac{\partial u}{\partial z} = -\frac{1}{\rho}\frac{\partial p}{\partial x} + \frac{\mu}{\rho}\left(\frac{\partial^2 u}{\partial x^2} + \frac{\partial^2 u}{\partial y^2} + \frac{\partial^2 u}{\partial z^2}\right) + g_x \tag{3}$$

$$\frac{\partial v}{\partial t} + u\frac{\partial v}{\partial x} + v\frac{\partial v}{\partial y} + w\frac{\partial v}{\partial x} = -\frac{1}{\rho}\frac{\partial p}{\partial y} + \frac{\mu}{\rho}\left(\frac{\partial^2 v}{\partial x^2} + \frac{\partial^2 v}{\partial y^2} + \frac{\partial^2 v}{\partial z^2}\right) + g_y \tag{4}$$

$$\frac{\partial w}{\partial t} + u\frac{\partial w}{\partial x} + v\frac{\partial w}{\partial y} + w\frac{\partial w}{\partial z} = -\frac{1}{\rho}\frac{\partial p}{\partial z} + \frac{\mu}{\rho}\left(\frac{\partial^2 w}{\partial x^2} + \frac{\partial^2 w}{\partial y^2} + \frac{\partial^2 w}{\partial z^2}\right) + g_z \tag{5}$$

For the momentum conservation, ANSYS Fluent manual [27] proposes the Eqs. (6), (7) and (8) for simplified systems (2D axisymmetric systems) for axial and radial directions, respectively:

$$\begin{aligned}\frac{\partial}{\partial t}(\rho v_x) + \frac{1}{r}\frac{\partial}{\partial x}(r\rho v_x v_x) + \frac{1}{r}\frac{\partial}{\partial x}(r\rho v_r v_x) = -\frac{\partial p}{\partial x} + \frac{1}{r}\frac{\partial}{\partial x}\left[r\mu\left(2\frac{\partial v_x}{\partial x} - \frac{2}{3}\left(\nabla \cdot \vec{v}\right)\right)\right] \\ + \frac{1}{r}\frac{\partial}{\partial x}\left[r\mu\left(\frac{\partial v_x}{\partial r} + \frac{\partial v_r}{\partial x}\right)\right] + F_x\end{aligned} \tag{6}$$

and

$$\frac{\partial}{\partial t}(\rho v_r)+\frac{1}{r}\frac{\partial}{\partial x}(r\rho v_x v_r)+\frac{1}{r}\frac{\partial}{\partial x}(r\rho v_r v_r)=-\frac{\partial p}{\partial r}+\frac{1}{r}\frac{\partial}{\partial x}\left[r\mu\left(\frac{\partial v_r}{\partial x}+\frac{\partial v_x}{\partial r}\right)\right]$$
$$+\frac{1}{r}\frac{\partial}{\partial r}\left[r\mu\left(2\frac{\partial v_r}{\partial r}-\frac{2}{3}\left(\nabla\cdot\vec{v}\right)\right)\right]-2\mu\frac{V_r}{r^2}+\frac{2\mu}{3r}\left(\nabla\cdot\vec{v}\right)+\rho\frac{v_z^2}{r}+F_r \tag{7}$$

where

$$\nabla\cdot\vec{v}=\frac{\partial v_x}{\partial x}+\frac{\partial v_r}{\partial r}+\frac{v_r}{r} \tag{8}$$

Eq. (9) is the expression of the energy equation. The first term on the left-hand side is called the transient term; the second to the fourth terms are called advection terms, and the terms on the right-hand side are diffusion terms [26, 27]. Energy equation related temperature to time and space coordinates. With the energy equation, obtaining the temperature evolution and distribution in a fluid field becomes possible:

$$\frac{\partial T}{\partial t}+u\frac{\partial T}{\partial x}+v\frac{\partial T}{\partial y}\,w\frac{\partial T}{\partial z}=\frac{\lambda}{\rho C_p}\left(\frac{\partial^2 T}{\partial x^2}+\frac{\partial^2 T}{\partial y^2}+\frac{\partial^2 T}{\partial z^2}\right) \tag{9}$$

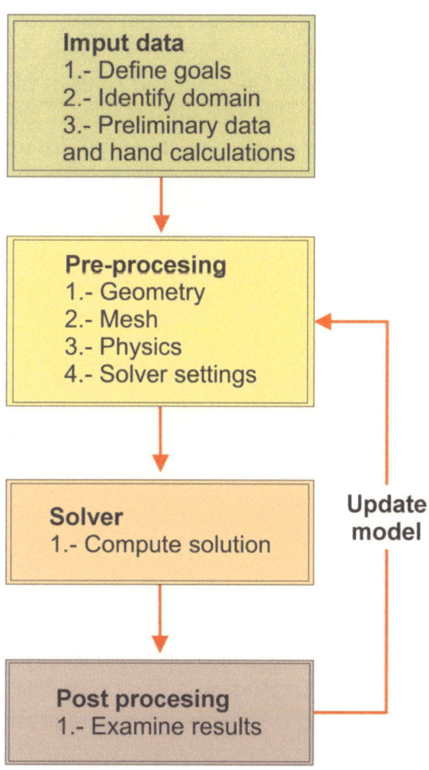

Figure 3.
ANSYS-Fluent CFD calculations process flow.

The ANSYS-Fluent manual [27] includes an alternative equation for calculating energy conservation which will not be included to avoid duplication of the same equation, but the reader is referred to the manual to study further if needed. If it is introduced a universal variable ϕ, a generic form of the governing equations could be written [26] as shown in Eq. (10),

$$\frac{\partial(\rho\phi)}{\partial t}+\frac{\partial(\rho u\phi)}{\partial x}+\frac{\partial(\rho v\phi)}{\partial y}+\frac{\partial(\rho w\phi)}{\partial z}=\Gamma\left(\frac{\partial^2\phi}{\partial x^2}+\frac{\partial^2\phi}{\partial y^2}+\frac{\partial^2\phi}{\partial z^2}\right) \tag{10}$$

where Γ is the diffusion coefficient. If ϕ = 1, Eq. (10) is the continuity equation. If ϕ = Xi, and Γ = D, Eq. (10) transforms to the component equation. If ϕ = u, v or w, Γ = μ, and the pressure term and gravity term are represented by source terms, equation (a10) becomes the Navier-Stokes equation. If ϕ = T, and Γ = λ/Cp, Eq. (10) turns out to be the energy equation. It is omitted the discussion of conservation of species and the effects of forces on the body because they are not being included nor resolved in this work. To close this section, ANSYS CFD solvers are based on the finite volume method. Domain is discretized into a finite set of control volumes. General conservation (transport) equations for mass, momentum, energy, species, etc. are solved on a set of control volumes for each one of the discretized volumes.

When one is working with a software such as ANSYS-Fluent [19] and other CFD programs, they obey a work structure that has to be followed with each calculation. The first step would be to do “manual” calculations to obtain the first estimates that will be used as input data for the next stage. The next step is called preprocessing, and it consists of using the input data to define an initial geometry, which is developed using CAD tools. This initial geometry is subjected to a discretization process where it is divided into small volumes to which the laws and rules of physics are assigned in the software settings, and it is established which equations are going to be solved in the next stage. The following step is the resolution stage, where numerical calculations are carried out using the computer and the program to reach a numerical solution for the governing equations, also known as the conservation equations. Finally, the last stage is called postprocessing, and in this stage, graphical tools are used, which build diagrams and graphs that help to better analyze the data and obtain conclusions from the results available in the numerical solution. **Figure 3** displays this framework followed by CFD simulations.

5. Computational details

5.1 Hydrodynamics simulation

Different geometries were modeled using PTC CREO from which the helical shape was selected for this work. The finite volume method (FVM) was used through the ANSYS Fluent software package to solve the governing equations of the CFD methodology. The fluid domain preparation was done with the SpaceClaim module included in ANSYS. The mesh preparation was done with the Mesh module included in ANSYS. A 0.1 cm hybrid mesh with 10 inflation layers in the area closest to the pipe wall was used. Additional mesh details are described in the mesh independence section in later sections. Preliminary model efficiency tests were performed in laminar flow. The hydrodynamic study was performed in turbulent flow regime using three different viscous models such as k-epsilon STD, k-omega STD, and k-omega SST. The hydrodynamic analysis was performed at different speeds; for the preliminary tests, wider speed ranges were used, ranging from 0.001 m/s (laminar flow) to 4 m/s (turbulent flow). For practical purposes, the results table displays only few laminar flow results since this work is interested more in the turbulent flow results. For the hydrodynamic analysis, the speeds that guarantee turbulent flow were used, ranging from 0.5 to 4 m/s. Once the numerical model was solved, postprocessing was carried out using ANSYS tools to obtain graphs from the numerical results for the hydrodynamic analysis.

5.2 Turbulent flow model

Turbulence flow problems require additional mathematical models to be considered in the simulation process. The first turbulence models accounted for the "eddy viscosity" or turbulent viscosity, since it reflects an apparent increase in viscosity caused by small-scale chaotic motions in a fluid. The concept of turbulence viscosity plays a central role in Reynolds Averaged Navier Stokes (RANS) models. Other approaches do not rely extensively on the turbulent viscosity concept [28–30]. **Figure 4** displays a map of the turbulence models included in ANSYS-Fluent.

The k-ε model performs suitably away from walls, in the main flow. However, it has issues in the boundary layer zone, particularly with low Reynolds numbers. A significant development in CFD was brought forward by the development of k-ω model that replaced the transport equation for ε with a specific rate of turbulence dissipation, ω [27].

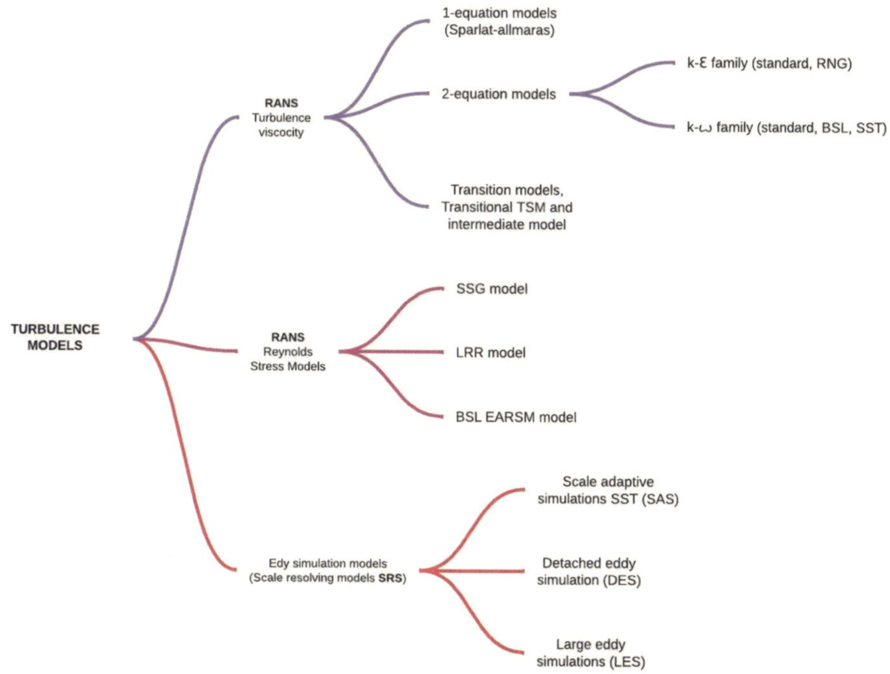

Figure 4.
Turbulence models summary map.

The standard k-ε and k-ω (empirical) models are based on transport equations for the turbulent kinetic energy (k), where ε = dissipation rate and ω = specific dissipation. The k-ε model relates Reynolds stresses to average flow properties. In this work, it is considered that the best turbulence method for this experiment was k-ω SST model. This method takes into account turbulent shear stress transport in the definition of turbulent viscosity. These features make the k-ω SST model more accurate and reliable for a broader class of flows (e.g., flows with adverse pressure gradients, airfoils, transonic shock waves), and it is only applicable to wall-bounded flows.

The Reynolds-averaged Navier-Stokes (RANS) equations have the same general form as the instantaneous (time-averaged or ensemble) Navier-Stokes equations. Other turbulence models such as the Reynolds Stress Model (RSM) closes these equations by solving transport equations for the Reynolds stresses, together with an equation for the dissipation rate. In contrast, unsteady RANS models are employed in the boundary layer, within Eddy Simulation (also known as DES which means Detached Eddy Simulation).

5.3 Validation notes before the numerical computations

The best-known validation for a numerical model is to validate the theoretical results with an experiment and reach low error levels, where the difference between both the theoretical and experimental results is considered the error. Other validation techniques involve comparative methodologies where the theoretical results obtained in a computational project that lacks experimental procedures are compared to historical experiments that were successfully validated, and by analogy, with the proper justifications, the newer numerical results are accepted if the differences are small enough. One of the more accepted alternatives to validate a methodology or procedure is the sensitivity analysis [31–34]. The CFD methodology proposed in this work has been widely used in other helical coil problems except for the design particularities and the viscous model used [35–40]. The k-omega models are recognized as a significant development in CFD that replaced the transport equation for k-epsilon models with a specific rate of turbulence dissipation, ω. The Shear Stress Transport model (SST) smoothly transitions from the k-ω model near the wall to the k-ε model in the main flow; thus, better results are expected for a helical shape pipeline [28, 41–44]. In the pages ahead, the k-omega SST viscous model is compared with two different viscous models to establish the differences in the results, taking into account that an improved performance by the mathematical model for k-omega SST is expected, as explained in the literature.

6. Reactor geometry definition

6.1 Reactor geometry modeling

A chemical reactor is any device used to carry out a chemical reaction, that is, one where molecular compounds are transformed into other molecular compounds [25]. It is the piece of equipment in which the conversion of raw materials to the desired product takes place and is therefore the only irreplaceable component of the process [23–25]. The geometry is the first feature to consider in the design and modeling of a chemical reactor, and in many cases, the geometry shape will determine the type of reactor operationally. Within the form factors of chemical reactors, in this section, the first geometry designs were developed in PTC CREO 10.0 [45] and ANSYS SpaceClaim [19]. These geometries were similar to different types of reactors previously selected from the literature but with customizations for our experiment based on standard laboratory-level measurements.

Among the different shapes, one can observe it is considered a curvy trajectory, which benefits the residence time of water in the reactor. Also, the continuous flow reactor type was preferred, which may be helpful with problems such as stagnation, vortex formation, or excessive turbulent kinetic energy.

6.2 Helical reactor geometry advantages

The literature mentions helical pipes may have improved performance in heat transfer, mass transfer, and fluid mixing, compared to straight pipes. Many practical advantages have been reported, such as compactness, easy fabrication, and high efficiency in heat.

Because of these advantages, helical pipes and derived reactors are often used in industrial applications, including power generation, chemical industry, nuclear industry, heat recovery systems, and refrigeration [1–4]. Some of these advantages were important in our selection of the reactor shape for this work.

Helical pipes generate intense Dean-type vortices. Evaluation of the friction factor in such geometries is recommended to assess the trade-offs between increased transfer efficiency and the associated specific energy requirement. Similar to other types of curved pipes and ducts, the most important dimensionless number used to characterize flow in helical pipes is the Dean number [3–5].

6.3 A brief view to different shapes of reactor geometries

Continuous flow reactors can provide greater control over mixing, improved thermal management, and safer access to hazardous chemicals over conventional batch reactors, among other types, and also, there are a great variety of reactor geometric shapes [1, 2]. An example of continuous reactors reported the possibility to further enhance the catalytic effect, especially in turbulent flow regimes. They also indicate that catalytic effects in continuous reactors depend on the geometry of the reactor and the “aging” of its components. Smaller diameter reactors with high surface-to-volume (S/V) ratios show greater catalytic effects due to increased molecular interaction between reactants and the reactor wall [3]. Other researchers chose open U-profiles. UV reactors in ballast water treatment systems often require the use of multiple high-power UV lamps in one reactor. To achieve the bactericidal effect, during operation, the UV reactor must be controlled so that the UV lamp dose remains constant and complies with the set standard dose [5, 7]. A straight tube with either circular or rectangular sections have been proposed for fully developed flow conditions looking for specific conditions in the nozzle exit to match their physical model. Other designers mentioned this shape allowed them to take advantage of the symmetry in the x-y plane at the centerline, so half of the channel was modeled [7]. Coil-based reactors are frequently structured as helically wound tubes, which has several advantages: forming a helix from the coil makes the reactor significantly more compact, reducing the physical space occupied by the unit [6]. **Figure 5** displays some of the different geometries analysed for this work such as helical, zig zag, tubular, and compartments, where, the first one was selected for further developments.

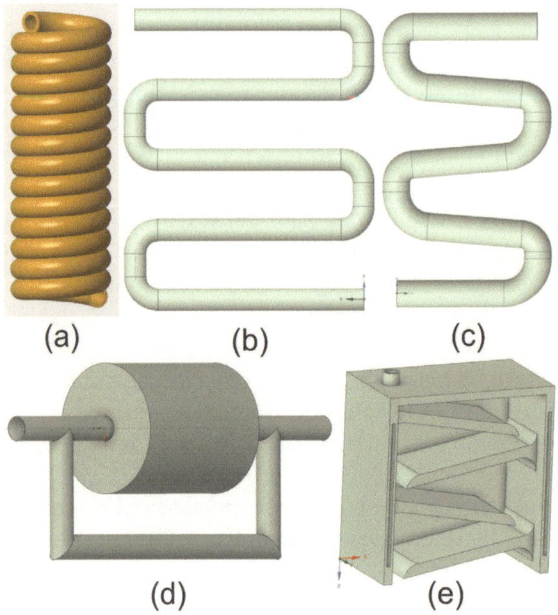

Figure 5.
Different geometries analyzed. (a) Helical or spiral, (b) zig zag 1, (c) zig zag 2, (d) tubular, and (e) compartments.

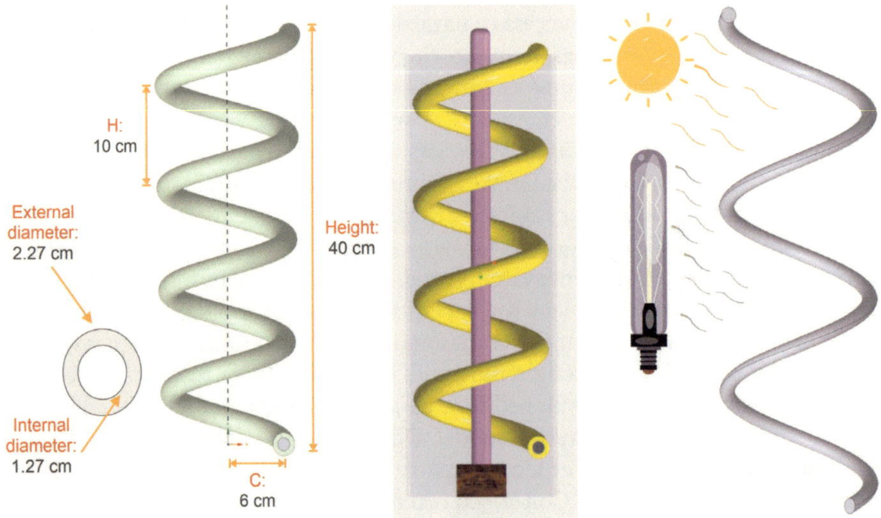

Figure 6.
Selected geometry and its potential use in photocatalytic reactors, (a) helical shaped pipeline, (b) scheme of helical pipeline shaped in a photocatalytic reactor using a centralized lamp, (c) scheme of a helical pipeline shaped photocatalytic reactor subject to sunlight and a lamp.

Item	ν (m/s)	Re_L	Dean (De)	Class
1	0.01	1588.3	458.5	Laminar
2	0.02	3176.6	917.0	Transition
3	0.03	4764.9	1375.5	Turbulent
4	0.04	6353.2	1834.0	Turbulent
5	0.05	7941.5	2292.5	Turbulent
6	0.06	9529.7	2751.0	Turbulent
7	0.07	11118.0	3209.5	Turbulent
8	0.08	12706.3	3668.0	Turbulent
9	0.09	14294.6	4126.5	Turbulent
10	0.1	15882.9	4585.0	Turbulent
11	0.5	79414.6	22925.0	Turbulent
12	1	158829.2	45850.0	Turbulent
13	1.5	238243.7	68775.0	Turbulent
14	2	317658.3	91700.0	Turbulent
15	3	476487.5	137550.1	Turbulent

Table 1.
Results for Reynolds number, Dean number, and flow regime classification.

7. Reactor simulation preprocessing with ANSYS Fluent

A reactor geometry design developed in ANSYS Spaceclaim, as shown in **Figure 6**, was chosen. This reactor is a helical-type reactor with a 40 cm vertical drop; this favors the continuity of the fluid flow within the reactor, and the inner diameter is half an inch, but since centimeters were the units selected, the dimensions are $\varnothing_{int}$ 1.27 cm and $\varnothing_{ext}$ 2.27 cm. The presence of four turns with a distance between them of H = 10 cm can be observed, as well as the main axis, which is C = 6 cm. **Figure 6(a)** shows the geometry that was selected for this experiment. It is a spring-shaped or helical geometry through which the fluid circulates. This pipe may be incorporated in different reactor models. For example, in **Figure 6(b)**, a section of a photocatalytic reactor that may have a central lamp source is displayed. In **Figure 6(c)**, a second and a third model, which may include a lamp source located at the sides of the pipe, is displayed, while the sunlight as the source for the photocatalytic reactor is also suggested.

The Reynolds and Dean number calculations are shown in **Table 1**. Most results are reported for inlet speeds between 0.01 and 3 m/s, as summarized in **Table 1**. For the figures within the results discussion section, only velocities that ensure a clearly turbulent regime—well separated from transition—were retained for the CFD analyses. The next parameters were constant in all cases of the different Reynolds number calculations: fluid density (ρ) was 998.2 kg/m^3, fluid viscosity (μ) was 0.001003 kg/m s, pipe diameter (d) was 0.0127 m, and pipe length was 0.1596 m.

The Reynolds number (Re) is a dimensionless value used in fluid mechanics and transport phenomena to characterize the motion of a fluid. It is calculated using a mathematical formula that relates the density, viscosity, velocity, and typical dimension of a flow. Eq. (11) is used to calculate the Re number as shown next:

$$Re = \frac{\rho * V * L}{\mu} \tag{11}$$

The value of the Reynolds number (Re) indicates whether the flow is laminar or turbulent: (a) if the Reynolds number is less than 2300, the flow will be laminar and (b) if the Reynolds number is greater than 4000, the flow will be turbulent. The laminar flow for the proposed geometry was found at flow velocities of maximum 0.01 m/s. The transition flow was observed between 0.02 and 0.03 m/s. And the turbulent flow was observed at flow velocities over 0.03 m/s. The Re number values for the different flow velocities are displayed in **Table 1**. The Dean number (De) is a dimensionless number representing the ratio between the viscosity force acting on a fluid flowing through a curved pipe and the centrifugal force. Eq. (12) is used to calculate the De number as shown next:

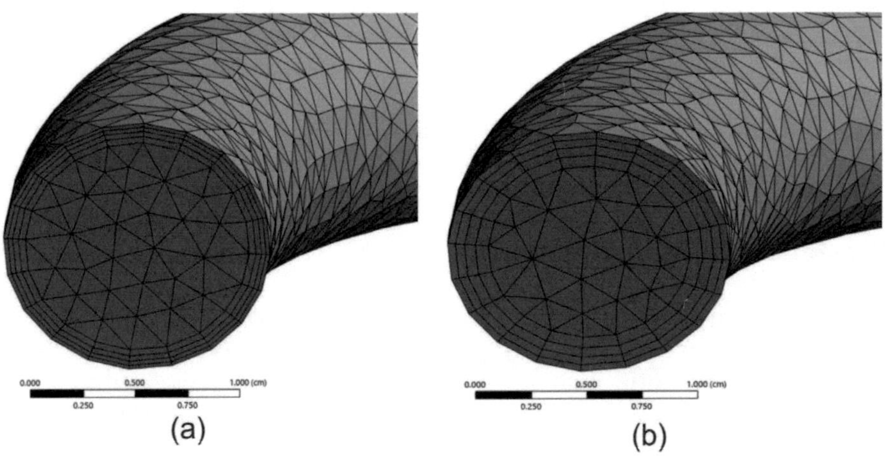

Figure 7.
Different mesh sizes: (a) extra fine (0.25 cm), (b) fine (0.3 cm).

Mesh class	Element size	Element quantity	Orthogonal quality	Aspect rate	Residuals	Cycles
Coarse-1	0.7	56,693	0.9385 ± 0.5995	6.892 ± 1.217	1×10^{-6}	359
Coarse-2	0.6	160,541	0.9600 ± 0.1506	12.700 ± 1.218	1×10^{-6}	479
Medium	0.5	160,541	0.9600 ± 0.1506	12.697 ± 1.218	1×10^{-6}	311
Fine-1	0.4	167,609	0.9779 ± 0.1252	14.419 ± 1.199	1×10^{-6}	325
Fine-2	0.3	183,434	0.9994 ± 0.1516	17.108 ± 1.169	1×10^{-6}	367
Extra fine	0.25	209,405	0.9913 ± 0.2003	16.136 ± 1.171	1×10^{-6}	418

Table 2.
Different mesh sizes were tested for mesh independence test.

$$De = Re\sqrt{\frac{L}{D_h}} \tag{12}$$

The Dean number displays high values even for some of the Re values within the range of laminar flow for the proposed geometry; for example, it is the case for flow velocities of 0.01 m/s. The transition and the turbulent flow also display high values of the De number, which is expected. The De number values for the different flow velocities are displayed in **Table 1**.

Figure 8.
Velocity vector plots arranged for the different positions along the pipeline for the selected velocities.

8. Helical pipeline hydrodynamics

In **Figure 7(a)**, one can find the mesh model that was used in this helical shaped pipe. The mesh numerical results are displayed in **Table 2**, and the reader may see in the last row that the selected mesh model was classified as extra fine mesh within the scope of this work.

The details related to the different mesh classifications can be seen in **Table 2**. For the selected extra-fine mesh, it contains more than 209,000 elements and also

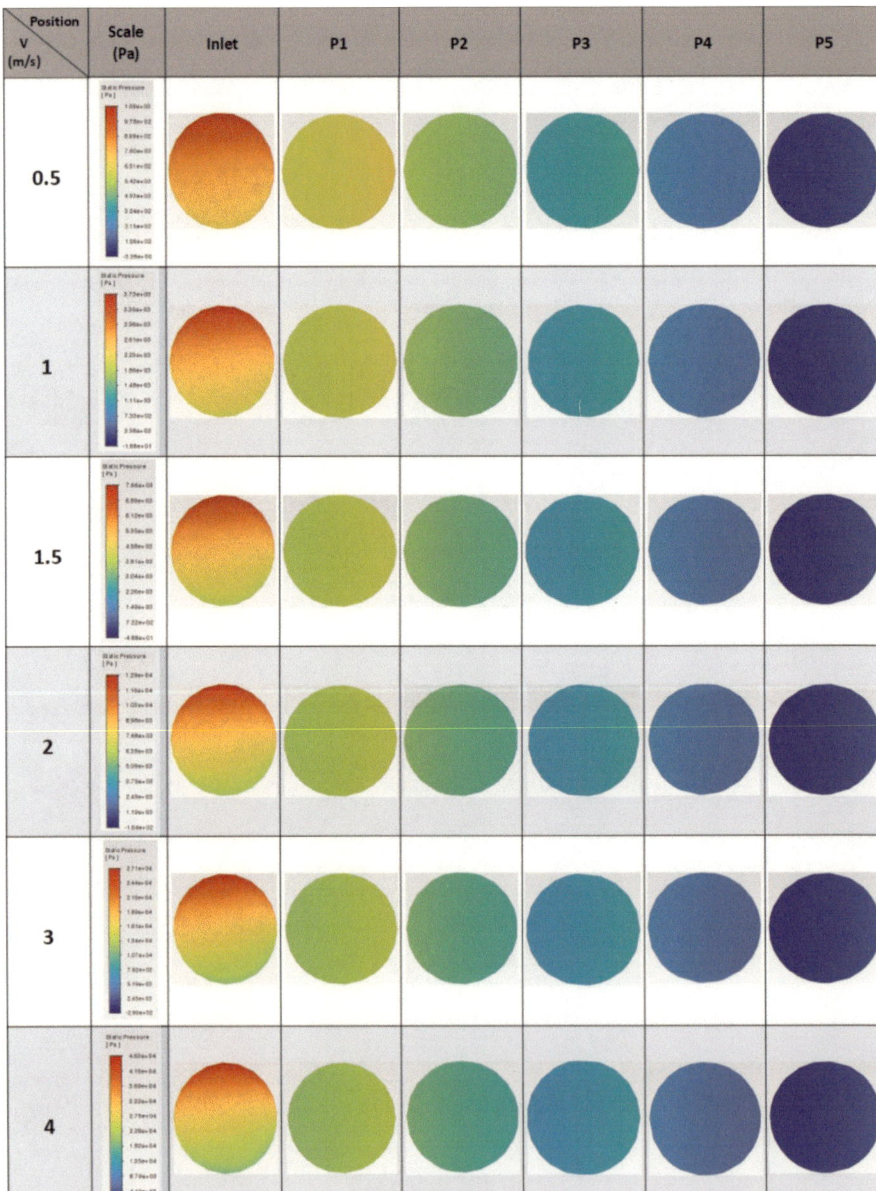

Figure 9.
Pressure contour plots arranged for the different positions along the pipeline for the selected velocities.

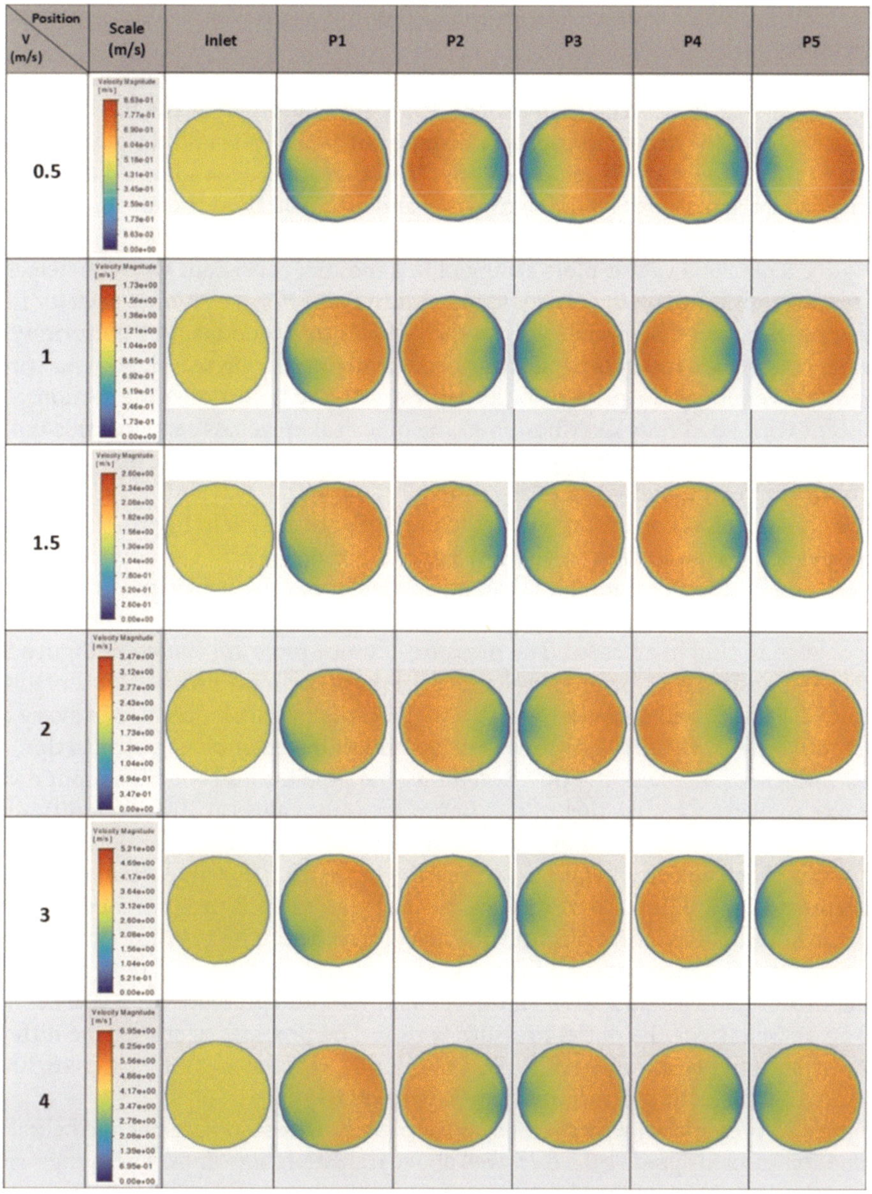

Figure 10.
Velocity contour plots arranged for the different positions along the pipeline for the selected velocities.

contains some inflation layers that are being used to have smaller elements for a better calculation of the fluid near to the pipe wall. Then, this mesh model is the one selected as the best results from the mesh independence tests.

For visualization purposes only, five virtual planes were created in the pipeline, which represent the positions where the different contour plots were obtained, and they would be labeled P1, P2, P3, P4, and P5, respectively. The five planes distributed evenly through the pipe length as well as the inlet and outlet were used to create fluid velocity, pressure, and turbulent kinetic energy contour plots displayed in the next pages. The results for the different velocities were summarized in a mosaic that displays the velocity contour plots for the different plane positions throughout the pipe length.

All of the velocity contours display a near zero slim region of fluid located next to the pipe wall. The flow velocity in general has a growing trend as the flow advances through the pipeline due to the vertical disposition of the tube and the gravity effects. The higher-flow velocity values are colored in orange, and depending on the input velocity, which has values from 0.5 to 4 m/s and after the fluid travels through the system, the flow velocity increases with values around 0.69–5.56 m/s.

Figure 8 presents vector plots arranged in a mosaic, consistent with the velocity and pressure contour layouts. The vectors indicate a uniform sectional velocity field, consistent with well-distributed flow in the pipe. Across sections, this uniformity supports continuous transport and strengthens with increasing speed, with no vortices observed. Overall, vector plots show a uniform velocity field across the section, consistent with good flow distribution in the pipe. Taken across sections, this uniformity indicates continuous through-pipeline transport, which strengthens as velocity magnitude increases, and vortices are generally absent.

The contour plots corresponding to the pressure are shown in **Figure 9**. These plots are arranged in mosaic, alike to **Figures 8** and **10**, organized by velocity (rows) and cross section location (columns). It can be observed that the behavior of the pressure tends to become uniform; if we analyze the different sections, one may notice the color change is mild in all cases. The pressure contour plots are shown in **Figure 9** for the selected positions at different velocities. In these sections, it can be observed that the color that is related to the pressure value presents mild color change in every case; in fact, only the inlet section displays a more noticeable color change due to this, which is the section where the flow is fed into the pipeline and contains input data.

If one analyzes the variation of maximum pressure values at P1 for the different velocities, an increasing trend in correlation with the velocity may be noticeable, that means that as the velocity increases so does the pressure. The values span from 651 Pa for 0.5 m/s to 18,200 Pa for 4 m/s. Since the outlet pressure is programmed to be zero, the pressure drop does correlate with the velocity as well. The pressure drop may be measured by difference between the pressure values at the inlet and outlet. Then, if the inlet pressure is around 1090 Pa for 0.5 m/s, then the pressure drop will be 1090 Pa as well, since the outlet pressure is zero. The pressure drops for the different positions have the next values: 3730 Pa, 7660 Pa, 12,900 Pa, 27,100 Pa and 46,300 Pa for 1 m/s, 1.5 m/s, 2 m/s, 3 m/s and 4 m/s, respectively.

Figure 10 presents velocity contours at successive cross-sections along the helical pipeline for selected speeds ≥0.5 m/s, well above transition flow limit. The orange–red bands reflect curvature-induced asymmetry across the sections. The sectional velocity maximum shifts from left to right (and vice versa) along the helix. An analogous left–right migration is seen for low-speed regions (blue), tracking the helical geometry.

Turbulence kinetic energy (TKE) is the mean kinetic energy per unit mass associated with eddies in turbulent flow. Physically, the turbulence kinetic energy is characterized by measured root-mean-square (RMS) velocity fluctuations. In the Reynolds-averaged Navier Stokes equations, the turbulence kinetic energy can be calculated based on the closure method, that is, a turbulence model. A very small effect due to swirl may benefit the chemical reactions during the helical reactor operation, but such effects may affect the flow continuity, so the designer needs to be careful with these effects.

Figure 11 shows the contour plots for the turbulence kinetic energy in the different positions along the tube in a mosaic arrangement for the different velocities. The

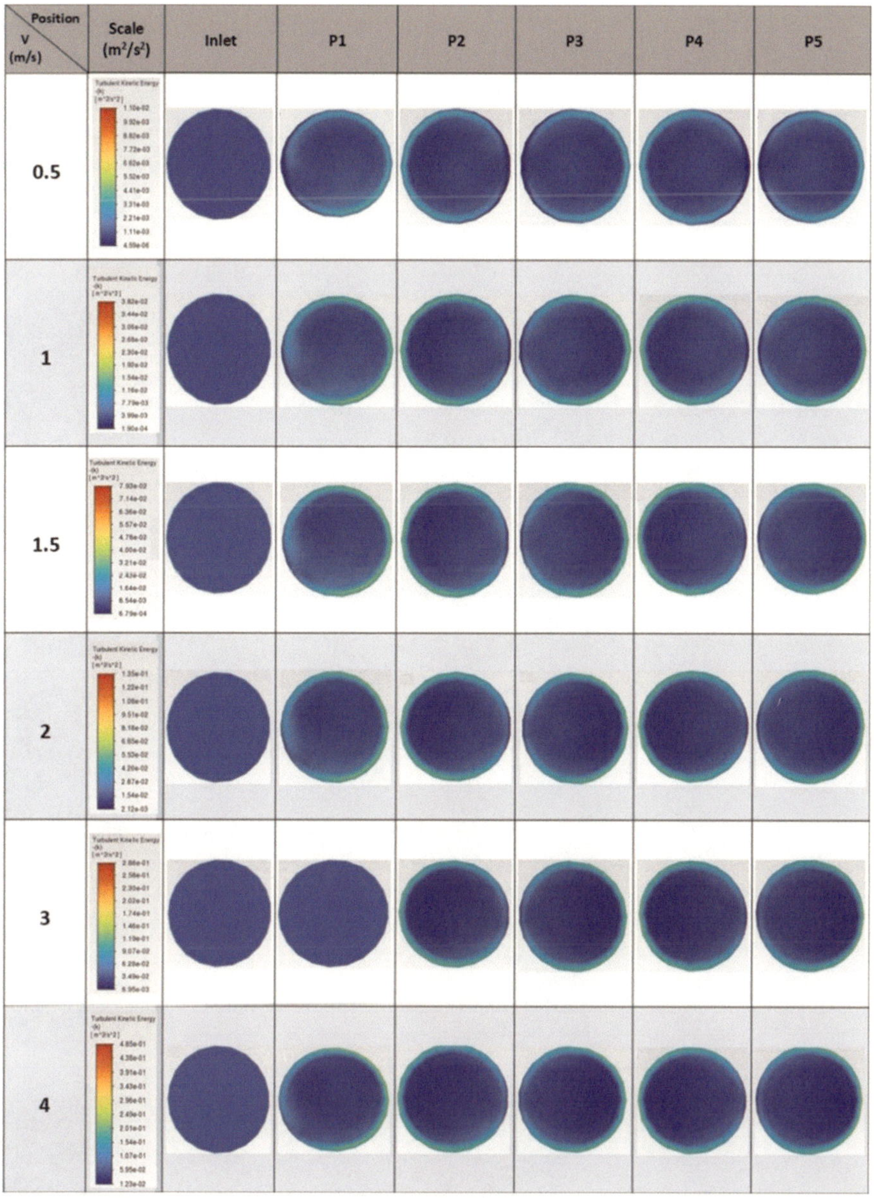

Figure 11.
Turbulence kinetic energy contour plots arranged for the different positions along the pipeline for the selected velocities. All calculations were made using k-omega SST viscous model.

intensity of the TKE higher values in the proposed model is shown in green color and may be observed near the pipe wall boundary in a portion near ¾ of the circumference. One may notice that TKE locates by the external side of the tube curvature due to the centrifugal forces. The maximum TKE values for each velocity are as follows: for 0.5 m/s TKE is around 0.004 m^2/s^2, for 1 m/s TKE is around 0.015 m^2/s^2, for 1.5 m/s TKE is around 0.032 m^2/s^2, for 2 m/s TKE is around 0.055 m^2/s^2, for 3 m/s TKE is around 0.190 m^2/s^2, and for 4 m/s TKE is around 0.201 m^2/s^2. If one notes that 0.5 m/s is working at turbulent regime, it may be better to avoid TKE values too far from this

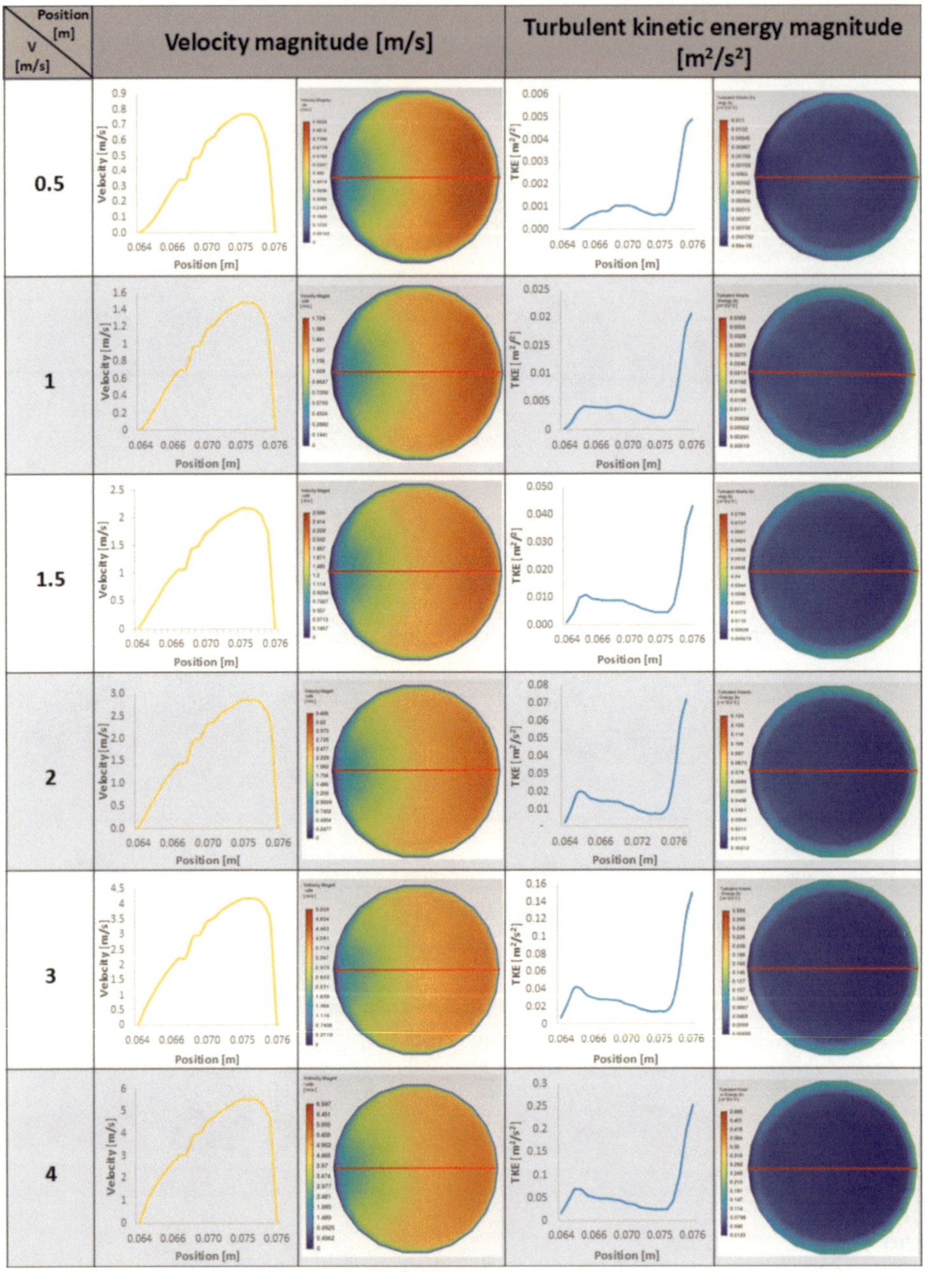

Figure 12.
X-Y graphics for velocity and TKE profiles. Also, the corresponding velocity and TKE contour plots arranged for the selected velocities. All graphs and contour plots correspond to the 2 m/s velocity.

value. On the other hand, the vectors plots do not show any continuity issues at first look for any of the cases displayed for velocities up to 4 m/s. In general, it can be observed that two colors predominate: green and blue.

This means that on one side of the tube, a higher energy value can be seen, but the green color is displayed in a small area more like a slim fringe near the pipe boundary wall. Alternatively, on the rest of the tube section, the TKE value is low, nearly zero,

DOI: http://dx.doi.org/10.5772/intechopen.1011998

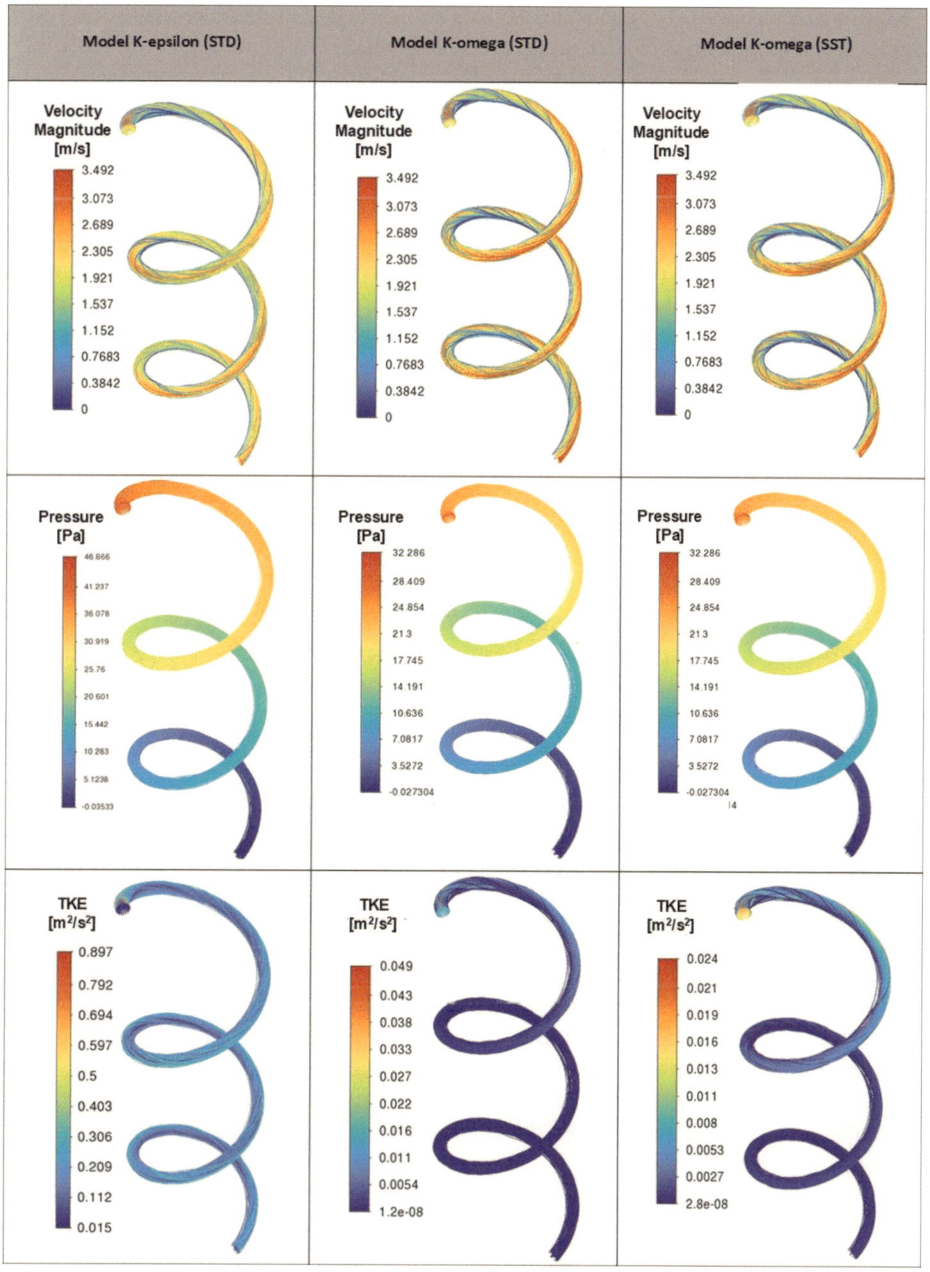

Figure 13.
Path line dynamic simulations for velocity magnitude, pressure, and TKE. Whole pipeline view for V = 2 m/s, for the selected viscous models.

which is colored in blue. In this figure, it can be seen that the green color appears on a rather small region and that the blue color predominates in the rest of the tube, while the continuity seems to work fine according to the other graphs; then one may conclude TKE results do not affect the hydrodynamics functionality of this reactor, but it does provide a tolerable amount of TKE to agitate the flow through the helical-shaped pipeline.

Figure 12 presents X-Y graphs illustrating the variation in velocity magnitude and turbulence kinetic energy (TKE) along a red line, which is a diametral route to build velocity and TKE profiles for the different flow velocities (0.5, 1, 2, 3, and 4 m/s). The analysis is performed through a central line, indicated in red, located on each of the pipe sections contour plots for the outlet location of each of the flow simulations. The corresponding velocity and TKE numerical values are extracted and used to build each X-Y graph. The behavior of the velocity magnitude, as a function of the position along the central line, reveals a progressive increase in velocity if the profile is analyzed from left to right, and the evolution in colors can be seen from blue to orange, respectively. However, it should be noted that the effect of viscous forces indicates a velocity of zero in the pipe boundary wall layer along the circumference of the tube. On the other hand, the evolution of the turbulence kinetic energy (TKE) in the outlet section may be observed in the X-Y graph, which represents a profile obtained from the observed values in the red line displayed in the corresponding TKE contour plot (also, for each velocity value). This graph shows an increase only in the external region of the tube curve. In the rest of the section, the TKE value is very low, almost zero. This effect could be related with the fact that the TKE has a direct relationship with the geometry of the helical pipeline, so the external side of the curve is the one that accumulates the highest TKE values (in green), and the rest of the section has very low TKE values (in blue).

Figure 13 displays pathline-based, whole-length views of the helical pipe, assembled from representative frames of dynamic simulations to verify through-pipe continuity. Shown are velocity magnitude, pressure, and TKE for three turbulence models (k–ε, k–ω Standard, k–ω SST) at an inlet speed of 2 m/s. **Figure 13** was organized in the form of a mosaic organizing in the columns the selected viscous models, and in the rows three different parameters such as the velocity magnitude, pressure and TKE, were considered. The flow velocity of 2 m/s was chosen because among the values selected for turbulent flow, this is the average velocity value within the values of interest for the flow selected in this work. In a previous section in this document are explained different reasons why the k-omega SST model is considered the most appropriate for this helical reactor, according to the literature and preliminary considerations for this project. In the case of the first row in **Figure 13**, the plots displayed correspond to the velocity magnitude for the selected viscous models using a starting 2 m/s as inlet velocity. In general, it can be observed that the velocity magnitude behavior is consistent among the three different viscous model variants presented in this figure, which means that the overall effects are similar but with slight differences in the numerical values. The highest values of the velocity magnitude correlate with the effect of the centrifugal forces related to the helical geometry of the reactor. The maximum velocity magnitude values show that the k-epsilon STD model finds slightly lower values of speed than both of the k-omega models. The k-omega STD model and the k-omega SST model predict very similar results in the case of the velocity magnitude values. In the second row of **Figure 13**, the images corresponding to the pressure for the selected viscous models can be observed, and they are also screenshots from the calculated path line dynamic simulations. In general, for the case of pressure, the three models also show consistent results or behavior; however, the values found are different. The k-epsilon STD model overestimates the pressure values, while the k-omega STD and k-omega SST models find similar values. These three images show the pressure change relates directly with the geometric shape of the helical pipeline since the colors related with the pressure values change in accord with each turn of the helical shape.

For the case of TKE, the graphs also show an overall consistent behavior, but it is difficult to observe the effect of the viscous forces because of the natural 3D disposition of helical pipeline. However, at this point, the reader may have a good idea of the TKE behavior based on the prior analysis corresponding to the different sections for TKE in the contour plots. TKE screenshots give a good idea of the green color, which represents the higher TKE values in the simulations, which also correspond with the contour plot observations, but at some point, the green color is hidden due to the helical disposition. However, the viscous models differ in their result values for TKE noticeably, if the differences are analyzed proportionally. The model k-epsilon STD overestimates TKE by 18x if compared with k-omega STD and 37x if compared with k-omega SST. From another perspective, k-omega STD overestimates TKE by 2x if compared with k-omega SST. Based on arguments explained in Section 5.3, viscous model k-omega SST is the best option for this work.

9. Conclusions

A helical shaped pipeline is proposed to be used in a reactor for applications in water treatment by means of photocatalysis. The hydrodynamics of the pipeline were studied using CFD numerical simulations. Calculations were made using six different fluid velocities, 0.5, 1.0, 1.5, 2.0 3.0 and 4.0 m/s. The mesh model was carefully selected, and it was denominated as extra-fine according to the classification described within this text and tested to probe mesh independence. The hydrodynamics of the model worked fine without any continuity or swirl problems. The k-omega SST turbulence model was used in each one of the selected velocity variants for the models reviewed in these simulations, resulting in acceptable TKE levels that do not interfere in and acceptable flow performance. The TKE was resolved with two other additional viscous models to establish the relevance of the viscous model in the numerical solution. The literature indicates the k-omega SST mathematical model is stronger than the k-epsilon and k-omega STD for systems like the one proposed in this work, so its results are accepted as the best choice for our model. It was noticed that the selected geometry shaped as a helical pipeline has a lot of influence in the results obtained in this experiment since the geometry curves represent the more noticeable feature for the flow continuity. The proposed geometry hydrodynamics performs well if the pipeline is used in a photocatalytic reactor as intended due to a good flow continuity and acceptable TKE values that may benefit the chemical reaction.

Acknowledgements

Thanks to the Laboratory for Advanced Computing and Emerging Technologies (LCATE) at FCFA-UJED for computational resources. Thanks to the Academic Group UJED-CA-129 for valuable discussions.

Conflict of interest

Authors state that this research was completed without any conflicts of interest related with the funding to develop the present work.

Appendix and nomenclature

$\vec{a}$ = *acceleration* $\left(\frac{m}{s^2}, \frac{ft}{s^2}\right)$

α = *permeability* $\left(\frac{L}{m^2.h.atm}\right), \left(\frac{ft^3}{ft^2.h.\left(\frac{lb_f}{ft^2}\right)}\right)$

c = *concentration* $\left(\frac{mass}{volume}, \frac{moles}{volumen}\right)$

d = *diameter* (m, ft)

D = *fluid domain*

C_D = *drag coefficient*

C_p = *heat capacity at constant pressure* $\left(\frac{J}{kg.k}, \frac{Btu}{lb_m}, ^\circ F\right)$

E = *total energy* (J, kJ, Cal, Btu)

E_r = *internal energy* $\left[h - \frac{p}{\rho} + \frac{1}{2}\left(v_r^2 - u_r^2\right)\right]$

$\vec{F}$ = *force vector* (N, lb)

g = *gravity* $\left(9.81\frac{m}{s^2}\right)$

H = *total enthalpy* $\left(\frac{Energy}{Mass}, \frac{Energy}{Mole}\right)$

H_r = *total enthalpy* $\left(H_r = E_r + \frac{p}{\rho}\right)$

κ = *kinetic energy per unit mass* $\left(\frac{J}{kg}\right), \left(\frac{Btu}{lb_m}\right)$

k = *turbulence kinetic energy*

p = *pressure* $\left(Pa, Ib_f/in^2, atm, mmHg, \frac{lb_f}{ft^2}\right)$

ρ = *density* $\left(\frac{kg}{m^3}\right), \left(\frac{lb_m}{ft^3}\right)$

r = *radius* (m, ft)

S = *tot.entropy* $\left(\frac{J}{K}\right), \left(\frac{J}{kmol-k}\right), \left(\frac{Btu}{lb_m mol-^\circ F}\right)$

T = *temperature* $(K, ^\circ C, ^\circ R, ^\circ F)$

t = *time* (s)

V = *volume* (m^3, ft^3)

$\vec{v_r}$ = *overall velocity vector* $\left(\frac{m}{s}\right), \left(\frac{ft}{s}\right)$

ω = *specific dissipation rate* (S^{-1})

$\vec{v}$ = *absolute velocity*

$\vec{\omega}$ = *angular velocity*

μ_t = *eddy viscosity* $\left(\mu_t = \frac{\rho k}{\omega}\right)$

ϕ = *equivalence ratio (dimensionless)*

μ = *dynamic viscosity* $\left(cP, \frac{Pa}{s}, \frac{lb_m}{ft-s}\right)$

$\vec{u_r}$ = *moving frame rate*

$\vec{v_t}$ = *intertial reference frame rate*

u, v, w = *velocity magnitude* $\left(\frac{m}{s}\right), \left(\frac{ft}{s}\right)$

q = *heat flux* $\left(\frac{W}{m^2}, \frac{Btu}{h}\right)$

$2\vec{\omega}.\vec{v_r}$ = *Coriolis acceleration*

$\vec{\omega}.\vec{\omega}.\vec{r}$ =	*Centripetal acceleration*
$\lambda =$	*molecular mean free path* (m, nm, ft)
$\bar{\bar{\tau}}_r =$	*stress tensor* $\left(Pa, \frac{lb_f}{ft^2}\right)$
$\nabla =$	*Nabla* (*Del*); *gradient or divergence*
$x_w =$	$X \rightarrow variable$ $w \rightarrow water$
$\phi =$	*equivalence ratio* (*dimensionless*) (*Phi*)
$\Gamma =$	*diffusion coeficient*

Author details

Diana Barraza-Jiménez[1*], Derian Manuel Lerma-Mancinas[1], Sandra Iliana Torres-Herrera[2], Hugo Iván Flores-Hidalgo[3], Elva Marcela Coria-Quiñones[4], Sayda Dinorah Coria-Quiñones[5] and Manuel Alberto Flores-Hidalgo[1*]

1 Laboratory for Advanced Computing and Emerging Technologies (LCATE), Faculty of Forestry and Environmental Sciences, Juarez University of Durango State, Durango, Mexico

2 Faculty of Forestry and Environmental Sciences, Juarez University of Durango State, Durango, Mexico

3 TecNM/Jiménez Institute of Technology, Jiménez Chih, Mexico

4 TecNM/Durango Institute of Technology, Durango, Mexico

5 CINVESTAV Querétaro Unit, Querétaro, Mexico

*Address all correspondence to: dianabarraza@ujed.mx; manuel.flores@ujed.mx

References

[1] Liu J, Li D, Chen H, et al. Timing the first emergence and disappearance of global water scarcity. Nature Communications. 2024;**15**:7129. DOI: 10.1038/s41467-024-51302-z

[2] Jones ER, Bierkens MFP, van Vliet MTH. Current and future global water scarcity intensifies when accounting for surface water quality. Nature Climate Change. 2024;**14**: 629-635. DOI: 10.1038/s41558-024-02007-0

[3] Gu W, Wang F, Siebert S, et al. The asymmetric impacts of international agricultural trade on water use scarcity, inequality and inequity. Nature Water. 2024;**2**:324-336. DOI: 10.1038/s44221-024-00224-78

[4] Wolkeba FT, Mekonnen MM, Brauman KA, et al. Indicator metrics and temporal aggregations introduce ambiguities in water scarcity estimates. Scientific Reports. 2024;**14**:15182. DOI: 10.1038/s41598-024-65155-5

[5] Veldkamp T, Wada Y, Aerts J, et al. Water scarcity hotspots travel downstream due to human interventions in the 20th and 21st century. Nature Communications. 2017;**8**:15697. DOI: 10.1038/ncomms15697

[6] He C, Liu Z, Wu J, et al. Future global urban water scarcity and potential solutions. Nature Communications. 2021;**12**:4667. DOI: 10.1038/s41467-021-25026-3

[7] Dolan F, Lamontagne J, Link R, et al. Evaluating the economic impact of water scarcity in a changing world. Nature Communications. 2021;**12**:1915. DOI: 10.1038/s41467-021-22194-0

[8] Song S, Li Q, Leslie G, Shen Y. Water treatment methods in heavy metals removal during photovoltaic modules recycling: A review. Resources Conservation and Recycling. 2024;**208**: 107701. DOI: 10.1016/j.resconrec.2024.107701

[9] Wichman CJ. The unequal burdens of water scarcity. Nature Water. 2023;**1**: 26-27. DOI: 10.1038/s44221-022-00016-x

[10] Zhang Z, Shan Y, Zhao D, et al. City level water withdrawal and scarcity accounts of China. Scientific Data. 2024; **11**:449. DOI: 10.1038/s41597-024-03115-4

[11] Emile R, Clammer JR, Jayaswal P, et al. Addressing water scarcity in developing country contexts: A socio-cultural approach. Humanities and Social Sciences Communications. 2022;**9**:144. DOI: 10.1057/s41599-022-01140-5

[12] Sustainable Development. The 17 Goals. Available from: https://sdgs.un.org/goals

[13] Secretary-General Warns Two Thirds of Global Population Could Face Water-Stressed Conditions within Next Decade, in Message for International Forests Day. SG/SM/17610-ENV/DEV/1641-OBV/1589. 2016. Available from: https://press.un.org/en/2016/sgsm17610.doc.htm#:~:text=By%202025%2C%20nearly%201.8%20billion,carbon%20capture%20and%20storage%20systems

[14] Laauwen M, Nowicki S. Reinforcing feedbacks for sustainable implementation of rural drinking-water treatment technology. ACS ES&T Water. 2024;**4**(4):1763-1774. DOI: 10.1021/acsestwater.3c00779

[15] Ciawi Y, Khoiruddin K. Low-cost antibacterial ceramic water filters for

decentralized water treatment: Advances and practical applications. ACS Omega. 2024;**9**:12457-12477. DOI: 10.1021/acsomega.3c09311

[16] Baccour S, Goelema G, Kahil T, et al. La gestión de la calidad del agua podría reducir a la mitad la escasez de agua futura de manera rentable en la cuenca del río Perla. Nature Communications. 2024;**15**:5669. DOI: 10.1038/s41467-024-49929-z

[17] Liu W, Liu X, Yang H, Ciais P, Wada Y. Global water scarcity assessment incorporating green water in crop production. Water Resources Research. 2022;**58**:e2020WR028570, 1-19. DOI: 10.1029/2020WR028570

[18] Pacheco-Treviño S, Manzano-Camarillo MGF. Review of water scarcity assessments: Highlights of Mexico's water situation. Wiley Interdisciplinary Reviews Water. 2024; **11**(4):e1721-e1739. DOI: 10.1002/wat2.1721

[19] ANSYS Fluent®, Academic Research CFD (Computer Software; Version/Release 19.1). Distributed and Published by ANSYS Part of Synopsys; 2019

[20] Ruiz-Soto A, Barraza-Jiménez D, Hurtado-Macias A, Torres-Herrera SI, Ríos-Orozco CO, López-Guzmán M, et al. Hydrodynamic Analysis on a Photocatalytic Reactor Using ANSYS Fluent®. London, UK: IntechOpen eBooks; 2020. DOI: 10.5772/intechopen.89782

[21] Barraza-Jiménez D, Ruiz-Soto A, Torres-Herrera SI, Coria-Quiñones EM, Olvera-Corral RA, Romero-Soto DJ, et al. Radiative Transference Equation Algorithm as an ANSYS® User-Defined Function for Solar Technology Applications. London, UK: IntechOpen eBooks; 2019. DOI: 10.5772/intechopen.88767

[22] López-Guzmán M, Flores-Hidalgo MA, Reynoso-Cuevas L. Electrocoagulation process: An approach to continuous processes, reactors design, pharmaceuticals removal, and hybrid systems—A review. Processes. 1831;**2021**:9. DOI: 10.3390/pr9101831

[23] Andriani G, De Liso BA, Pio G, Salzano E. Design of sustainable reactor based on key performance indicators. Chemical Engineering Science. 2024b; **285**:119591. DOI: 10.1016/j.ces.2023.119591

[24] Dübal S, Berkel LL, Debiagi P, Nicolai H, Faravelli T, Hasse C, et al. Chemical reactor network modeling in the context of solid fuel combustion under oxy-fuel atmospheres. Fuel. 2024; **364**:131096. DOI: 10.1016/j.fuel.2024.131096

[25] Froment GF, Bischoff KB, DeWilde J. Chemical Reactor Analysis and Design. 3rd ed. Hoboken, NJ, USA: John Wiley & Sons, Inc.; 2011. pp. 07030-5774. DOI: 10.1002/9781118136539. ISBN: 1118136535, 9781118136539

[26] Ji G, Zhang M, Lu Y, Dong J. The Basic Theory of CFD Governing Equations and the Numerical Solution Methods for Reactive Flows. London, UK: IntechOpen eBooks; 2023. DOI: 10.5772/intechopen.113253

[27] ANSYS, Inc. ANSYS Fluent Theory Guide. Canonsburg: ANSYS Inc.; 2020. Published for licensees' users by ANSYS® INC

[28] Gorman J, Bhattacharyya S, Cheng L, Abraham JP. Turbulence Models Commonly Used in CFD.

London, UK: IntechOpen eBooks; 2022. DOI: 10.5772/intechopen.99784

[29] Habeeb LJ, Al-Turaihi RSS. Turbulent Flow Simulations. London, UK: IntechOpen eBooks; 2021. DOI: 10.5772/intechopen.90251

[30] Ali H, Rasani MR, Harun Z, et al. Passive flow-field control using dimples for improved aerodynamic flow over a wing. Scientific Reports. 2024;**14**:12918. DOI: 10.1038/s41598-024-63638-z

[31] Tricomi L, Melchiori T, Chiaramonti D, Boulet M, Lavoie JM. Sensitivity Analysis and Accuracy of a CFD-TFM Approach to Bubbling Bed Using Pressure Drop Fluctuations. Frontiers in Bioengineering and Biotechnology Section: Chemical and Process Engineering. 2017;**5**:1-15. DOI: 10.3389/fbioe.2017.00038

[32] Yang LL, Padding JTJ, Kuipers JAMH. Modification of kinetic theory of granular flow for frictional spheres, part I: Two-fluid model derivation and numerical implementation. Chemical Engineering Science. 2016;**152**:767-782. DOI: 10.1016/j.ces.2016.06.015

[33] Nasr FM, Tano M, Sabharwall P, Azmy Y. Global sensitivity analysis using Gaussian process surrogate models & impact of thermophysical properties uncertainties on the steady-state operation of molten salt reactors. Annals of Nuclear Energy. 2025;**217**:111304. DOI: 10.1016/j.anucene.2025.111304

[34] Tian W. A review of sensitivity analysis methods in building energy analysis. Renewable and Sustainable Energy Reviews. 2013;**20**:411-419. DOI: 10.1016/j.rser.2012.12.014

[35] Ghazanfari V, Shadman MM, Mansourzade F, Amini Y. Numerical investigation of thermal-hydraulic performance enhancement in helical coil heat exchangers with twisted tube geometries. Case Studies in Thermal Engineering. 2024;**60**:104744. DOI: 10.1016/j.csite.2024.104744

[36] Di, G Sigalotti L, Alvarado-Rodríguez CE, Rendón O. Fluid flow in helically coiled pipes. Fluids. 2023;**8**(12): 308. DOI: 10.3390/fluids8120308

[37] Sun S, Liu J, Zhang W, Yi T. Frictional pressure drop for gas–liquid two-phase flow in coiled tubing. Energies. 2022;**15**(23):8969. DOI: 10.3390/en15238969

[38] Di Piazza I, Ciofalo M. Numerical prediction of turbulent flow and heat transfer in helically coiled pipes. International Journal of Thermal Sciences. 2009;**49**(4):653-663. DOI: 10.1016/j.ijthermalsci.2009.10.001

[39] Abushammala O, Hreiz R, Lemaître C, Favre É. Laminar flow friction factor in highly curved helical pipes: Numerical investigation, predictive correlation and experimental validation using a 3D-printed model. Chemical Engineering Science. 2019;**207**:1030-1039. DOI: 10.1016/j.ces.2019.07.018

[40] Wang J, Liu Y, Ding R. Analysis of heat transfer and flow characteristics of a helically coiled tube with twisted elliptical in a low Reynolds number flow. Processes. 2022;**10**(11):2229. DOI: 10.3390/pr10112229

[41] Wilcox DC. Multiscale model for turbulent flows. AIAA Journal. 1988; **26**(11):1311-1320. DOI: 10.2514/3.10042

[42] Jin Y, Kuznetsov AV. Multiscale modeling and simulation of turbulent flows in porous media. International Journal of Fluid Engineering. 2024;**1**(1): 010601-010615. DOI: 10.1063/5.0190279

[43] Menter FR. Two-equation eddy-viscosity turbulence models for engineering applications. AIAA Journal. 1994;**32**(8):1598-1605. DOI: 10.2514/3.12149

[44] Sparrow EM, Tong JCK, Abraham JP. Fluid flow in a system with separate laminar and turbulent zones. Numerical Heat Transfer, Part A: Applications. 2008;**53**(4):341-353. DOI: 10.1080/10407780701454162

[45] PTC Inc. PTC Creo Parametric (Version 10.0.0.0). Needham, MA: PTC Inc; 2024. [Computer Software]

Section 4

Resource Recovery from Wastewater Treatment Plants

Chapter 6

Innovative Utilisation of Wastewater Sludge: Integrating Treatment Processes with Resource Recovery for Sustainable Infrastructure

Kobe Samuel Mojapelo, Williams Kehinde Kupolati, Everardt Andre Burger, Julius Musyoka Ndambuki, Jacques Snyman, Emmanuel Rotimi Sadiku and Idowu David Ibrahim

Abstract

Sludge, a byproduct of wastewater treatment plants, poses significant environmental and operational challenges owing to increasing regulations and environmental concerns surrounding its disposal. Recent studies have investigated the conversion of organic semi-solid sludge into sustainable building materials, including bricks, glass ceramics, and lightweight aggregates, using advanced resource-recovery technologies. Anaerobic digestion and thermal hydrolysis are key pre-treatment processes that enhance the efficiency of sludge incineration for producing high-quality wastewater sludge ash (WSA). WSA can then be utilised as a supplementary cementitious material in concrete, asphalt, and bricks. Life cycle assessments demonstrate environmental benefits, including lower resource consumption, reduced CO_2 emissions, and material savings, among others. However, heavy metal leaching remains a significant concern; thus, the requirement for precise process control and regulatory compliance is also emphasised. This circular economy in wastewater treatment is a realistic application of the concept of a circular economy. However, successful treatment strategies depend on an integrated approach that combines monitoring and analytical technologies, environmental assessment, and supportive regulatory frameworks. Future research should focus on standardising treatment processes, developing quality assurance protocols, and improving market-adoption mechanisms.

Keywords: sludge treatment and management, resource recovery, wastewater sludge ash, circular economy, environmental sustainability

1. Introduction

1.1 Background and problem statement

The global wastewater treatment industry is facing an emerging challenge in effectively treating and disposing of the continually increasing volume of sludge produced as a byproduct of treatment operations. The current estimate of total global sewage sludge production is more than 53 million dry tonnes per year [1], with some studies reporting daily generation rates as high as 8.16 million tonnes [2]. Conventional sludge treatment methods mainly involve disposal through landfilling, ocean dumping, and agricultural use; however, these practices have become less environmentally friendly due to rising environmental concerns, stricter regulatory requirements, and increasing costs. Historically, the conventional method of defining sludge as a waste stream has consistently imposed environmental and financial obligations on municipalities and treatment plant operators [3, 4]. The costs associated with sludge disposal make up 50–60% of the total operating costs of a wastewater treatment plant, and they are increasing due to new regulations imposed for the disposal of used sludge [5]. Furthermore, traditional waste disposal raises environmental concerns due to greenhouse gas emissions from landfills, potential soil and groundwater contamination, and the loss of materials that could otherwise be recycled.

Recent regulatory developments have increased the urgency for innovative sludge management solutions. Such initiatives are driven by the European Union's Circular Economy Action Plan, with measures progressing in North America and Asia. Similarly, increasingly stringent environmental regulations worldwide are likely to shift from traditional practices to recycling-focused strategies in previously unconsidered regions [6]. These policy frameworks emphasise the importance of transforming waste streams into valuable resources and positioning sludge valorisation as a critical component of sustainable sludge management [7, 8]. This paper provides an overview of the state-of-the-art and future potential of innovative sludge utilisation technologies, particularly highlighting their integration with sustainable infrastructure development. The primary outcomes will be: (1) the analysis of sludge generation trends and characteristics according to the type of treatment; (2) advanced technologies for sludge treatment analysed together with changes in material properties; (3) performance in building applications analysed, especially for WSA materials; (4) environmental assessment of all its life cycle steps considered; (5) sectorial and regulatory contexts are also foreseen in order from the legal authorities' responsibility perspective, as well as regulatory compliance concerns; (6) in terms of market deployment bottlenecks and incentive options related to circular economy frameworks.

The circular economy has been working for years to transform sludge into a valuable resource. The stabilisation of sludge influences biogas production for renewable energy through processes such as anaerobic digestion. In thermal methods, sludge ash can be reused in construction applications, further promoting sustainable infrastructure [9, 10]. The current methods of construction and demolition (R&D) waste management can be redirected to address the increasing need for new eco-friendly construction materials without being dependent on virgin material resources [11].

This systematic review presents a literature review on sludge valorisation technologies and their contributions to sustainable infrastructure development. Central to the approach is a synthesis of treatment processes, material characteristics, environmental burdens and impacts, regulatory settings, and implementation barriers, aided by qualitative and quantitative tools. The remainder of this review is organised as follows, based on five main classifications: technical feasibility, environmental sustainability, economic viability, regulatory compliance, and practical implementation. The following classifications are briefly explained below.

i. Technical feasibility: This involves assessing the efficacy of the treatment process and optimising material properties and scalability [12].

ii. Environmental sustainability: This includes the evaluation of environmental impacts throughout the entire life cycle, encompassing resource depletion, emissions reduction, and environmental conservation [13, 14]. This principle also applies to economic analyses, where caution pertains to factors such as return on investment, market alignment, and potential economic entry barriers [15]. Guidance on environmental sustainability involves a comprehensive review of the total environmental impact, utilising appropriate life cycle assessment principles, and considering resource consumption, emissions abatement, and ecosystem protection.

iii. Economic viability: This includes economic analysis, market and cost-effective potential estimates, cost-benefit analyses, and the identification of economic barriers to deployment.

iv. Regulation compliance: Reviews existing and pending legal requirements, safety codes, and licensing needs in multiple areas. Furthermore, it reviews existing safety standards, licensing requirements, and the current regulatory landscape, which evolves across various regions for compliance.

v. Practical implementation: This is also referred to as operational implementation, which evaluates practicality, including infrastructure needs, operational complexity, and stakeholder support [16].

1.2 Conventional sludge management approaches

Conventional biosolid disposal methods involve dumping them into landfills, incineration, and application to land as fertilisers. Although these methods were previously considered sufficient, they are now being re-evaluated for their negative environmental impacts, such as greenhouse gas emissions and soil degradation resulting from land application [17, 18]. As wastewater sludge production increases, the requirements for purification processes have also increased owing to new regulations, such as bans on sea disposal and reductions in nutrient discharges.

These new sludge-to-energy processes are reflected in various emerging solid recovery fuel technologies, such as thermal treatments (anaerobic digestion and advanced digestion), which may help enhance resource recovery from sludge or more traditional production pathways, leading to biosolids or biochar end-products for land application purposes [19, 20]. These advancements have been essential in making WWTPs more adaptable and unique in adopting a circular economy model that not only treats water but also recovers resources from waste by-products to create sustainable solutions.

2. Wastewater sludge generation and characteristics

2.1 Wastewater sludge generation rate

Biological sludge is generated by the treatment of wastewater operations, and appropriate disposal practices are important for a sustainable plant. The amount of sludge produced is governed by several parameters, including the characteristics of the incoming sludge, the nature and quantity of treatment processes adopted, and plant operating conditions. A high volume of sludge is generated worldwide, and this amount continues to increase. The latest data also suggest that the world annually generates more than 53 million tonnes of sewage sludge on a dry weight basis [21], with some studies reporting daily generation rates as high as 8.16 million tonnes [22]. The growth trends reflect population growth, urbanisation and increasingly strict wastewater treatment standards imposed around the world. The activated sludge process is a widely practiced biological method of treating sludge and is one of the leading causes of generating sewage sludge. Over 70% of municipal WWTPs worldwide are equipped with this process, resulting in 0.3–0.6 kg DS per kg BOD removed from the sludge [1, 23]. Further advanced treatment schemes aimed at enhanced pollutant removal may generate even larger amounts of sludge, 20–30% more than conventional methods.

The system boundary of a wastewater treatment process, with the main subsystems responsible for material and energy flow, is shown in **Figure 1**. It includes a screen, grit chamber, and two parallel primary treatment routes, such as (i) an activated sludge process with a sequential batch reactor and (ii) a constructed wetland combined with an up flow anaerobic sludge blanket. These produce emissions and treat effluents. The sludge produced undergoes thickening and then moves into different stabilisation pathways, including anaerobic digestion, lime stabilisation, and advanced options such as anaerobic digestion with pre-treatment and lime stabilisation with electricity recovery, which contribute to reduced emissions and generate avoided products, including energy and fertilisers. **Figure 1** provides a clear overview of the wastewater treatment life cycle stages, linking environmental emissions, energy demand, and resource recovery, and is relevant for life cycle assessment (LCA) and sustainability evaluation.

Table 1 shows the physicochemical characteristics of the sludge at the Midvaal, Emfuleni, and Lesedi municipal WWTPs. The sludge is alkaline in nature, with a pH (6.5–8.2) that meets the requirements of biological treatment processes. The high Total Solids (TS) content (4.7 and 5.4%) suggests that a larger organic fraction is present, that is, volatile solids (VS) (3.2–4%), with a VS/TS ratio close to 0.7, and it can be used for aerobic processes as well as thermal recovery [25]. The increased total chemical oxygen demand (COD) (67,300–81,450 mg/L) and soluble COD (4,600–6,000 mg/L) enable the use of biodegradable loads that are sufficient to generate substantial energy. Depending on the type, the higher heating value of the sludge (11.3–14.2 MJ/kg) falls within an acceptable range for thermal valorisation, such as incineration at an optimal temperature of 600 °C, considering

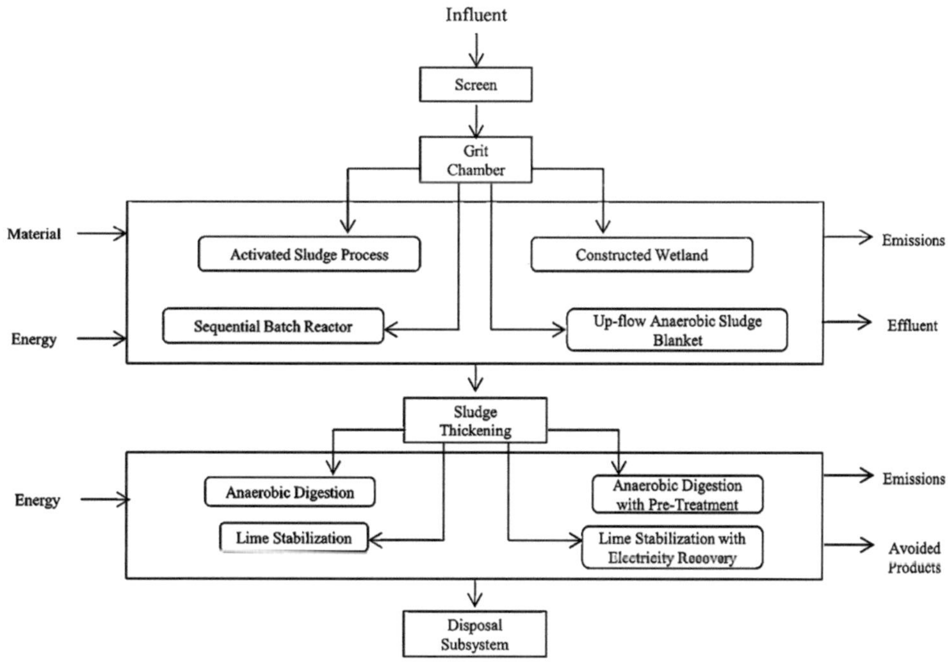

Figure 1.
System boundary of the wastewater treatment process [24].

Parameter	Unit	Value
pH		6.5–8.2
TS	% W/W	4.7–5.4
VS	% W/W	3.2-4
VS/TS		0.69–0.7
TCOD	mg/L	67,300–81,450
SCOD	mg/L	4600-6,000
Soluble sugar	mg/L	80
Soluble protein	mg/L	88
Bulk density	kg/L	1.26–2.56
Particle density	kg/L	2.4–2.56
Ash	g/kg	345-440
Higher heating value	MJ/kg	11.3–14.2

Table 1.
Characteristics of the WSA [25].

similar technologies. The value of ash content (expressed within the range of 345–440 g/kg) also suggests that the post-combustion phases generate waste that can likely be recovered and used in other cementitious materials. Both the bulk and particle densities fall within the ranges suitable for conventional sludge-handling operations. This study further highlights the importance of integrated sludge management in South African WWTPs, where energy recovery should at least be combined with resource recovery through WSA valorisation in construction processes, as indicated elsewhere, especially considering the findings regarding essential sludge characteristic levels.

The chemical composition of WSA, as presented in **Table 2**, is based on various studies and exhibits significant variations depending on the sludge source and treatment methods. Compared to the known range of pozzolanic activity, the SiO_2 content of this fume, which ranges from 12.4% to 49.5%, indicates significant pozzolanic activity. The current CaO content also overlaps, varying from 5.8% to 31.3%. The Al_2O_3 and Fe_2O_3 levels are moderately high, suggesting the potential to produce a durable binder mixture for most compositions. A few alkalis (Na_2O and K_2O) are generally low to moderate and have received attention because they influence durability. These data indicate that WSA can be classified as an environmentally friendly material suitable for construction applications.

2.2 Physicochemical properties and composition of WSA

The physicochemical characteristics of WSA are quite diverse because they depend on various factors, such as sludge origin, the wastewater treatment process applied, incineration conditions, and operating conditions. A knowledge gap exists regarding these properties, which are fundamental for evaluating the potential applications of WSA in construction.

2.2.1 Chemical composition and variability

The chemical composition of WSA primarily comprises SiO_2, Al_2O_3, Fe_2O_3, and CaO as the main oxides. The proportions of these oxides significantly influence the pozzolanic activity of the ash, which is ultimately responsible for its ability to react with calcium hydroxide in the presence of water and produce additional cementitious compounds. It has been found that the SiO_2 content in WSA ranges

Reference	SiO_2	Al_2O_3	Fe_2O_3	MgO	CaO	Na_2O	K_2O
[27]	12.4	14.9	7.4	2.6	31.3	-	-
[28]	39.03	15.13	17.05	1.93	5.80	0.43	0.62
[29]	28.34	12.44	18.60	3.21	10.63	7.43	1.92
[30]	26.95	8.13	8.21	4.64	21.80	1.93	1.31
[31]	21.97	18.62	12.93	2.6	21.73	0.056	5.83
[32]	49.5	12.9	12.6	2.17	6.35	3.43	
[33]	18.9	9.6	13.8	4.4	18.2	1.1	2.2
[34]	25.77	9.54	5.12	4.48	20.70	0.55	1.87
[35]	28.3	12.5	18.6	3.2	10.6	7.4	1.9
[36]	35.52	19.44	16.88	2.80	6.85	0.298	2.90

Table 2.
Chemical compositions of various WSAs [26].

from around 12.4% to 49.5%, and a higher SiO_2 content results in a more activated form [26]. The CaO content of the CCRs varied widely, ranging from 5.8 to 31.3%, depending on the chemical conditioning agents used for sludge treatment and lime-based stabilisation processes. High calcium levels can also enhance early strength; however, if not carefully controlled, expansion issues may arise. These usually contain 8–19% Al_2O_3 and 5–18% Fe_2O_3, such that dissolved aluminates and ferrites can provide long-term strength potential. The alkali content ($Na_2O + K_2O$) in WSA is generally lower than that of other SCMs (0.05%–7.4%). If not all reactions with the alkalis occur in the cement, higher alkali levels may lead to problems such as ASR in concrete.

2.2.2 Physical properties and characteristics

The physical characteristics of WSA have a significant influence on its performance in various applications. The specific surface area of silica (WSA) ranges from 300 to 1,000 m^2/g, which is much higher than that of OPC, typically 300–400 m^2/g, and also exceeds that of SF, with values between 15,000 and 25,000 m^2/g. This intermediate fineness enhances the pozzolanic reactivity and filler effects in cementitious systems [37].

It is typically coarser than cement, with d50 values normally ranging from 10 to 50 μm. The morphology of WSA particles is commonly irregular and angular, owing to thermal decomposition and sintering during incineration. This angular shape may result in a higher water demand for the concrete mix, while also improving the mechanical interlocking and bond strength. The bulk density of WSA ranges from 1.26 to 2.56 kg/L, while the particle density varies from approximately 2.4 to 2.56 kg/L, which are lower than those of typical mineral admixtures. This suggests that using WSA as a partial replacement for cement could lead to the development of lightweight concrete with additional benefits immediately after its demolition.

2.2.3 Thermal and durability properties

The mineralogical composition and thermal history of WSA during incineration are essential for understanding its thermal properties. The ash is generally thermally stable because it has been treated at high temperatures (850–1000 °C) during the incineration process. This thermal process promotes the growth of stable crystalline phases and removes organic residues that could compromise durability. The pozzolanic activity of WSA, measured by the strength activity index value, typically ranges from 75% to 95% at 28 days when compared with control cement mortars. This is mainly due to the presence of an amorphous phase of silica and alumina, which reacts with calcium hydroxide to generate more C–S–H gel, the primary binding phase in cementitious materials.

2.2.4 Heavy metal and environmentally conscious

Concerned issues. One of the significant concerns regarding the use of WSA is its contamination with heavy metals, which can accumulate during incineration. The typical heavy metals in WSA are zinc, copper, chromium, nickel, and lead, and their concentrations vary significantly depending on the industrial and domestic sources entering the wastewater treatment plant. According to recent literature, these high incineration temperatures can even immobilise most toxic substances by binding them into stable crystalline phases, thereby reducing their leachability. WSA is usually tested for extraction using methods such as the TCLP or EN 12457. Most research has demonstrated that concentrations of leachate from WSA are below hazardous waste levels, with even lower values often reported when the ash is encapsulated in cementitious matrices, providing additional physical entrapment and chemical bonding to immobilise the residues.

2.2.5 Quality control and standardisation

Because of the wide variability in WSA properties, strict QC/Standardisation protocols are mandatory. These include its chemical composition, physical characteristics, heavy metal content, and pozzolanic behaviour. Some investigations have proposed classifying WSA according to its chemical composition and performance properties, as is done for fly ash and other SCMs. The establishment of test standards and acceptance criteria for WSA is a key step in facilitating its application in construction. This includes limitations on specific chemical constituents, physical properties, and performance characteristics, such as strength, activity index, and durability requirements. Sludge is a heterogeneous material with a wide range of physicochemical properties that depend on its origin, treatment, and operational conditions. Such variability poses challenges and opportunities for valorisation applications that demand rigorous characterisation and quality control to ensure consistent material properties.

2.3 Conventional sludge management approaches

Typical sludge handling methods include proven techniques that have been used for many years. These include anaerobic digestion, land application, landfilling, and incineration, each with trade-offs that have led to their use in different countries under various regulatory regimes.

Anaerobic digestion is one of the most widely used biological processes for sludge stabilisation. This process involves the biodegradation of organic matter in the absence of oxygen, resulting in the production of gas (biogas) and a stable material (digestate). Recent studies have shown that anaerobic digestion can achieve volatile solid reductions of 40–60%, along with the generation of renewable energy in the form of biogas [38]. It performs best at mesophilic temperatures (35–40 °C) or within the thermophilic temperature range (50–60 ° C), the latter offering additional advantages such as faster reaction rates and improved pathogen destruction. However, it is a costly method and requires skilled labour to operate, which makes it challenging for most smaller treatment plants to adopt.

Land application of biosolids has been widely practiced as a beneficial reuse, especially in agriculture. It is cost-effective, and nutrient recycling by returning nutrients (N and P) bound or tied up in treated straw to the soil increases organic matter, thereby enhancing the soil's water-holding capacity. However, soil recycling has become increasingly regulated due to the accumulation of heavy metals, the spread of pathogens, and the emergence of pollutants such as pharmaceuticals and personal care products [39]. In recent years, the limits on heavy metal content have been tightened, and national laws have strengthened the control of soil and groundwater.

Landfilling remains the most popular destination for sludge disposal and treatment in many areas with limited treatment options. However, this method of waste disposal is becoming increasingly unsustainable in Japan, as the country has limited landfill space, increased disposal costs, and pressing environmental concerns. When sludge is landfilled, it generates biogas, and the methane within it can be released into the atmosphere as a greenhouse gas (GHG). Furthermore, landfill leachate can pollute groundwater and surface water resources [40]. In light of the growing restrictions on landfilling organic waste implemented in many developed countries, there is an increasing demand for alternative treatment methods for organic materials.

Incineration can achieve a high-volume reduction (up to 90%) and generate heat for energy recovery. New incineration facilities destroy all pathogens and detoxify organics at temperatures > 850 °C, and wastewater sludge ash (WSA) is suitable for construction work. However, this alternative has a low possibility due to high costs, and more air pollution control facilities, including those for dust, acid gas, and organic trace components, are necessary [41]. The energy balance of incineration can be positive if the sludge has a high calorific value, which typically indicates effective dewatering and pre-treatment.

Recent developments in integrated technologies for conventional sludge minimisation and valorisation aim to increase resource recovery by effectively combining various sludge treatment methods to reduce their environmental impact. For example, thermal hydrolysis pretreatment can boost anaerobic digestion efficiency by up to 50% and improve the dewaterability of the final digestate [42]. Similarly, a combination of anaerobic digestion followed by digestate incineration can help maximise energy recovery and volume reduction, generating recyclable, sterile ash.

The selection of appropriate traditional treatment alternatives will vary based on criteria such as sludge quality, local regulations, available equipment, financial resources, and public opinion. Currently, favoured systems combine several conventional techniques to optimise efficiency and resource recovery while minimising environmental effects.

2.4 Sludge generation in wastewater treatment plants as supplementary cementitious material

Biological and chemical processes in modern wastewater treatment systems inevitably generate sludge. This necessitates a thorough understanding of sludge production mechanisms and related influential factors to develop effective valorisation strategies and resource recovery solutions. Biological sludge generated in WWTPs originates from several sources, each with a distinct quality and proportion of total sludge. Primary sludge originates from the settling process, facilitating the further removal of contaminants and pollutants (e.g., organics and solids) such as dissolved chemicals, settleable solids, floating materials, and colloidal materials in raw sludge. It has a high percentage of organic matter. The second is sludge from biological treatment, which is primarily composed of microorganisms resulting from the fermentation of organic matter and the removal of nutrients.

More than 70% of municipal WWTPs worldwide operate based on the activated sludge process, which produces a significant amount of biological sludge during treatment operations, and periodic removal is necessary for optimal efficiency. The characteristics of secondary sludge are primarily influenced by operational conditions, including the age of the sludge, organic load rate (OLR), and nutrient content. The generation of sludge depends largely on the treatment technology, operating conditions, and properties of the influent. Generally, conventional activated sludge systems produce 0.3–0.6 kg of dry sludge per kg of biochemical oxygen demand (BOD) removed, whereas advanced treatment processes can yield 20–30% more sludge owing to improved contaminant removal mechanisms [23, 43]. These high production rates are notable when performing a rough calculation at the municipal level, recognising that large treatment plants can generate hundreds of tons of sludge annually.

Figure 2 shows the scanning electron micrographs (SEM) of the WSA pyrolysed at 600 °C obtained from the Polokwane Municipality. **Figure 2(a)** at 100,000x magnification, shows the very porous, lumpy surface structure with agglomerated fine particles, clearly indicating the incomplete decomposition of the organic mass and the formation of amorphous clusters on the surface due to heat treatment. In **Figure 2(b)** (magnification = 50,000x), well-defined, larger interconnected voids and crushed ash particles are observed, which may promote a higher surface area as well as increased roughness, conducive to pozzolanic activity.

3. Advanced treatment technologies

New trends in sludge treatment aim to promote a 'resource recovery' and product generation approach, adding value, where waste minimisation also leads to reduced environmental impact. These innovative technologies are expected to recover energy, nutrients, and resources from wastewater sludge to the greatest extent possible, thereby lowering disposal costs and environmental burdens.

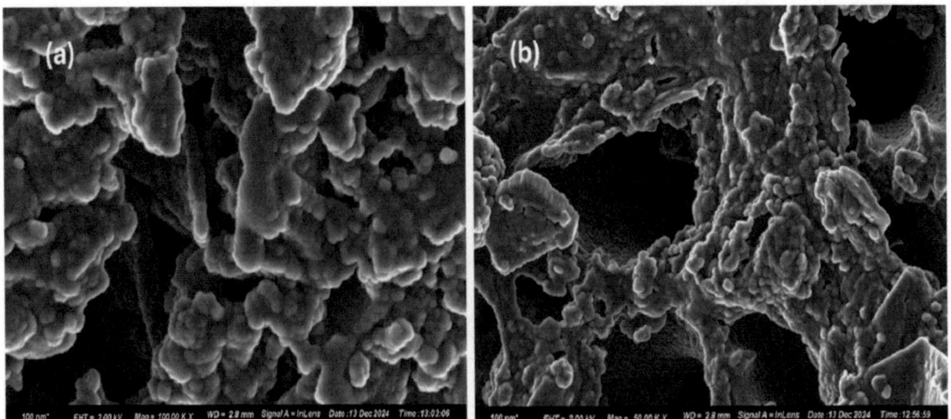

Figure 2.
SEM micrographs of WSA pyrolysed at 600 °C .

3.1 Thermal hydrolysis

Thermal hydrolysis is an improved pretreatment that utilises high pressure (6–10 bar) and temperature (150–180 °C) to break down complex organic matter in sludge, making it more biodegradable and easier to dewater. In this process, the sludge is heated in an autoclave for 20–30 minutes and then rapidly depressurised, which ruptures cell walls and releases intracellular organic matter. The benefit of thermal hydrolysis is that it significantly boosts the efficiency of subsequent anaerobic digestion. Pretreatment with thermal hydrolysis increases methane production by 40–50% relative to conventional mesophile digestion and also decreases the hydraulic retention time from 20–25 days to 12–15 days [44]. It also enhances the dewaterability of digested bio-solids, with a dry solids content of 25–35% for the AR reactor versus 18–22% for conventional digestion. Thermal hydrolysis also provides effective pathogen elimination, meeting Class A biosolid requirements and allowing for unrestricted land application. Vegetative bacteria, viruses, parasites, pharmaceuticals, and other emerging contaminants are essentially killed by high temperatures and pressures. Thermal Hydrolysis Process (THP) on Thickened Sludge (TS): Recent progress in the field of thermal hydrolysis has included improved control and reduced energy consumption, for instance, as demonstrated by Thermogram 5, presented by Obernberger et al. Modern heat recovery systems can recover over 85% of this thermal energy and, in conjunction with highly efficient anaerobic digestion, ensure a neutral or even positive energy balance in the long term.

3.2 Incineration and ash production

Incineration is the oldest process for reducing sludge volume and recovering energy. Currently, incinerators operate at temperatures exceeding 850 °C to ensure the complete elimination of organic materials, including pathogens and organic pollutants. This process reduces the volume by 85–90% and mass by 70–80%, making waste disposal significantly more convenient. Sophisticated sludge incineration techniques (e.g., fluidised bed reactors, multiple hearth furnaces, and rotary kilns) offer advantages for specific sludges and treatment purposes in terms of meeting treatment targets. The good heat and mass transfer capabilities of FB systems enable cold (>750–850 °C) yet effective combustion to be performed with a fair degree of temperature control. The primary product resulting from the treatment of sludge by aerobic incineration is WSA, which generally accounts for 10–15% of its dry matter. Its composition heavily depends on the features of the incoming sludge and incineration plant operating conditions. Still, it is typically rich in silica, alumina, and calcium oxide, offering pozzolanic activity suitable for construction purposes. In some countries, energy recovery from sludge incineration can be highly significant, producing net electricity when the dry solid content in sludge exceeds 25% in new plants. Heat recovery systems can produce steam for electric power generation and district heating, and state-of-the-art combined heat and power systems can achieve a total energy efficiency of 80–85%.

3.3 Pyrolysis

Pyrolysis is a thermal drying process that converts sludge into biochar (solid), bio-oil (liquid), and syngas (gas) by heating sludge in the absence of oxygen at temperatures ranging from 300 to 700 °C. The relative ratios of these products can be adjusted by temperature, heating rate, and residence time to optimise them for the desired end-use application. Maximum maturation occurs with slow pyrolysis (heating rates < 10 °C/min, at a temperature between 300 and 500 °C), which maximises biochar production (25–35%), making it suitable for use as a valuable adsorbent or soil conditioner, as well as fuel. Typically, the pyrolysis biochar of sludge (SPB) has a high surface area (300–800 m^2/g) and easily adsorbs heavy metals and organic contaminants. The goal is to produce bio-oil that can be upgraded into fuel or used as a feedstock in the chemical industry. However, the nitrogen and oxygen content of bio-oil obtained from sludge pyrolysis is generally high, and further upgrading steps are required for its application as a fuel. Pyrolysis enables syngas production, and the syngas (15–25% yield) contains hydrogen, carbon monoxide, and methane, which can be combusted for heat or electricity. New developments in pyrolysis include chemical upgrading to improve yield quality and symbiotic gasification to optimise energy recovery.

3.4 Gasification

Gasification is a thermochemical method for decomposing biomass at high temperatures (700–1000 °C), generating syngas as a potential source of clean and renewable energy, while utilising organic components in sludge using oxygen, steam, or carbon dioxide. This is performed under sub-stoichiometric conditions, allowing for complete combustion to be avoided and maximising the gas yield. The syngas from sludge gasification generally contains 15–25% hydrogen, 15–30% carbon monoxide, 5–15% methane, and 35–50% nitrogen and carbon dioxide. This gas can be used to generate heat and power, as well as in gas engines and turbines, because it is a syngas that can be further processed to produce hydrogen, methanol, or synthetic fuels through Fischer-Tropsch synthesis. Gasification offers advantages over incineration, including higher energy efficiency (75–85% compared to 65–75%), lower emissions of air pollutants, and the ability to produce a solid residue ('char') that can be utilised as a soil conditioner or construction material. It also enables precise control of heavy metal distribution, with most metals remaining solid-bound rather than volatilised. Advancements in the gasification process (e.g., plasma arc gasification) have facilitated treatment at very high temperatures (1000–1,500 °C), wherein all organic materials are destroyed, and the remaining constituents are vitrified into a stable, inert form. These developments enable the processing of heavily contaminated sludge and the generation of high-value-added syngas for chemical synthesis.

3.5 Alternative binder applications

The advanced processing of sewage sludge can generate secondary materials suitable for use as alternative binders in construction applications. Alkali-activated materials (AAMs) are a potential alternative technology that utilises industrial secondary resources, such as WSA, as raw materials activated through an alkaline solution to produce cementitious binders. WSA can serve as an aluminosilicate source in geopolymer systems activated with sodium or potassium hydroxide and silicate solutions. The geopolymer binders produced possess compressive strengths comparable to those of PC, in addition to enhanced chemical resistance and a lower

CO_2 footprint. Recent studies have reported the successful utilisation of WSA up to 50% (by weight) as a raw material in geopolymer mortars [45]. Magnesia-based cements are another option for binding, with WSA used as an additional component in the mixture. These systems offer rapid setting, high early strength gain, and excellent durability. The addition of WSA can enhance workability and reduce the environmental impact during the production of magnesia cement.

3.6 Aerated concrete applications

Autoclaved Aerated Concrete (AAC) is a lightweight construction material that can be processed to utilise WSA as a raw material effectively. It involves eliminating gas-forming bubbles during foaming, followed by hardening at 180– 220 °C through steam pressure after creating a cellular structure by stirring cement, lime, sand, and aluminium powder [46]. In the production of AAC, WSA can substitute up to 30% of fine sand by volume while maintaining acceptable mechanical and thermal properties. The WSA is responsible for the higher density and strength of the autoclaved matrix due to its finer particle size and pozzolanic behaviour. Recent studies have demonstrated that AAC with WSA can achieve a density of 400–600 kg/m^3 and compressive strength of above 3–5 MPa, meeting the international standard for structural AAC [47]. The inclusion of WSA in AAC also benefits environmental protection and waste valorisation and reduces the exploitation of natural sand. Material and method-based LCA studies indicate that WSA-containing AAC can lower its carbon footprint by 20–30% compared to traditional AAC manufacturing.

3.7 Lightweight aggregate production

Synthetic lightweight aggregates can be produced from WSA by pelletising and sintering, serving as a sustainable alternative to natural lightweight aggregates such as pumice or expanded clay. Production involves binding WSA with (typically) clay or cementitious binders, followed by pelletising and firing at temperatures between 1000 °C and 1,200 °C to achieve a product with suitable density and strength. Lightweight WSA aggregates (bulk density levels varying from 800 to 1200 kg/m^3) with crushing strengths of 2–8 MPa can be used in structural lightweight concrete applications. The cellular microstructure formed by sintering provides sound thermal insulation and low energy consumption, while also being recyclable, making it a desirable candidate for sustainable construction materials. Recently, the development of lightweight aggregates has been known to utilise a foaming agent to reduce firing temperatures and energy consumption, or to mix with other waste materials, such as fly ash or bottom ash, to improve properties and achieve cost savings in manufacturing. Furthermore, cold bonding techniques using cement or geopolymer binders have been devised for the production of low-temperature lightweight aggregates [48].

4. Resource recovery uses of ash from wastewater sludge

The paradigm of the circular economy has led to a transformational shift in how slime is perceived, moving it from waste to an invaluable resource. Innovative treatment techniques for WSA, which have been proven to be environmentally friendly, are utilised as materials for concrete. This study presents the broad application of WSA in various building materials, focusing on performance properties, optimisation methods, and ecological advantages.

4.1 The use of WSA in concrete

There has been growing interest in the utilisation of WSA as an alternative SCM in concrete. It is assumed that cement consumption will decrease while maintaining or increasing the concrete's properties. WSA can substitute Portland cement in concrete mixes, contributing to waste valorisation and the reduction of carbon dioxide emissions.

4.1.1 Mechanical properties and performance

Several studies have demonstrated that, with acceptable mechanical properties, the success rate of substituting WSA ranges from 5% to 20% (mass-based). Mojapelo and Kupolati [49] found that the compressive strengths of 5–10% WSA concrete after 28 days were comparable with or slightly higher than control values (85–105% strength activity indices). The optimum replacement level was between 10% and 15%, which showed the best synergy between the pozzolanic and filling effects. The pozzolanic reaction of WSA with calcium hydroxide, which is produced during cement hydration, contributes to long-term strength gains. This reaction begins and becomes significant after 7–14 days, with continued strength development beyond 28 days. Research indicates that concrete containing WSA achieves a compressive strength 10–15% higher at 90 days than similar data at 28 days or less, suggesting activation of the pozzolanic addition. The workability of WSA concrete is affected by the ash particle size distribution and surface area. The WSA is usually finer than cement, and it can increase the water demand by 5–15%, which must be considered in the mix design. However, its increased fineness is believed to decrease bleeding and improve particle packing, allowing for lower writing pressure, enhanced cohesion, and a higher quality of finish.

4.1.2 Durability performance

WSA as a fine aggregate has been studied in various exposure and concrete durability issues, and positive results have been reported in nearly all exposure cases. The pozzolanic reaction of WSA promotes the refinement of the pore structure, thereby reducing permeability and resisting aggressive substances. WC flow (WSA, 10–15 Vol.%). The results indicate that concrete with 10–15% WSA content can resist surface chloride penetration approximately 20–30% lower than the control mix, ensuring good performance against chloride attack. The sulphate attack on WSA-concrete is generally suitable, especially when the ash has a low sulphate content and a proper chemical composition. The generation of more C–S–H gel through pozzolanic reactions can benefit the sulphate resistance of the cement matrix. However, the assessment of WSA composition should be considered to prevent potential expansion issues related to delayed ettringite formation. The carbonation resistance of WSA concrete may be slightly reduced owing to the consumption of Ca by the pozzolanic reaction. However, the enhanced porosity may outweigh this, yielding carbonation rates similar to or higher than those of the concrete reference. Furthermore, adequate curing and minimum cover are becoming increasingly crucial for long-term behaviour.

4.1.3 Environmental considerations

Regarding environmental impact, waste can be valorised by WSA in concrete, but potential heavy metal leaching should also be considered. WSA added to concrete matrices exhibits very low leachability compared to raw ash, as evidenced by numerous extensive leaching studies, where most heavy metals are immobilised through chemical bonding and physical encapsulation. LCA studies indicate that using WSA in concrete can reduce the carbon footprint by 8–15% for replacements between 10% and 20%. This reduction results from both decreased cement emissions and waste diversion from landfills. The transportation distance of WSA and the energy required for ash utilisation can influence the overall environmental benefit.

4.2 Application of WSA in the asphalt

The use of WSA as a mineral filler for asphaltic mixtures would enable the use of waste materials, such as wastewater sludge ash, instead of traditional mineral fillers, such as limestone dust, creating an alternative outlet for this by-product. Using WSA as a filler can enhance pavement performance and contribute to waste management.

4.2.1 Filler properties and performance

WSA has properties that make it a suitable mineral filler in bituminous mixes. The particle size distribution of WSA generally conforms to the standard for mineral fillers, with 70–90% of the particles passing through a 75 μm sieve. Due to their limited shape and large specific surface area, WSA particles are expected to provide a better stiffness factor and interlocking effect with aggregates. Azarhoosh and Asadi [50] found that the Marshall stability and flow values of asphalt mixtures with WSA as a partial substitute filler (25–50% by weight) were higher than those of the control mixtures. The most advantageous substitution was observed at a level of 30–40%, when the physical and chemical effects combined to improve efficiency without compromising practicality. WSA-modified asphalt mixtures appear to demonstrate favourable fatigue responses, with some studies recording fatigue life improvements of 15–25% compared to conventional mixes. This improvement can be explained by the improved interaction between the binder and filler and the possible chemical bond with the molecules of the asphalt binder.

4.2.2 Rutting resistance and high-temperature performance

The use of WSA improved the rutting and high-temperature behaviours of asphalt mixtures. The surface texture and angularity of WSA increase the stiffness of the mix and resistance against rutting. The incorporation of WSA can significantly improve the anti-rutting and high-temperature resistance of the mixture. The angular particles and greater surface area per unit weight contribute to a stiffer mix and enhanced resistance to deformation. Several experimental studies have demonstrated that WSA-treated RAP mixtures achieve 10–20% higher dynamic stability in wheel rutting performance than control mixtures. Additionally, as a modifier in asphalt binders, WSA can raise the high-temperature continuous performance grade by approximately 6–12 °C, according to some studies.

4.2.3 Moisture susceptibility and durability

The moisture sensitivity of WSA-modified asphalt mixtures is a significant issue, likely due to the high water absorption capacity of ash compared to that of conventional mineral fillers. Research reports have identified that a better-quality airfield mix, formulated with anti-stripping agents, may help address moisture-related problems in airfield pavements. Nadupuru and Jagtap [51] reported that WSA mixes, when treated with the appropriate anti-stripping agent (asphalt), can achieve TSR values ranging from 85 to 90%, which are characteristic of moisture-sensitive mixes. If the WSA moisture content, binder content, and compaction are manipulated, good moisture resistance can be attained. Very few long-term durability studies on WSA-modified asphalt pavements are available; however, the results obtained so far indicate that well-designed mixtures should perform at least as well as conventional HMA. Field demonstrations and accelerated testing programs have been conducted to evaluate the long-term performance of such eco-friendly pavements.

4.3 Utilisation of WSA in bricks

Another potential application of WSA in the brickmaking industry is its ability to reduce consumer dependence on virgin clay and provide an environmentally acceptable medium for ash disposal. WSA plays a dual role as a raw material and flux during the preparation of brick products.

4.3.1 Manufacturing process and optimisation

WSA can be utilised in brick-making through various methods, including the direct addition of WSA into clay mixes, using WSA as a fluxing agent to reduce firing temperatures, and incorporating WSA into lightweight brick formulations. The most appropriate integration method depends on the properties of the WSA and should be customised to meet the specific production requirements of the bricks. Adding WSA directly to clay mixes at replacement rates between 10 and 30% by mass is common, but larger quantities can affect workability and green strength. WSA is finely textured and acts as a plasticiser in clay mixes; its chemical composition may induce flux effects during firing. It has been confirmed that adding WSA can reduce the firing temperature by 50–100 °C without compromising brick properties. This fluxing agent, WSA, is particularly beneficial for saving energy during the brick production process. The WSA was phase-pure because the alkali and alkaline earth oxides interact to facilitate liquid-phase formation by realising sintering and densification at lower temperatures. These factors can reduce the firing energy of bricks by 10–20% and enhance their strength.

4.3.2 Mechanical properties and quality

During manufacturing and sintering, the mechanical properties of WSA-based bricks were comparable to those of regular bricks. Compressive strengths ranging from 15 MPa to 25 MPa are typically achievable with WSA contents of up to approximately 20% in structural applications, thereby satisfying most building code standards. The water absorption of bricks containing WSA is typically satisfactory, ranging from 12% to 18%, depending on the amount of WSA and the firing conditions. The porosity of WSA can cause a slight increase in water absorption, which can be controlled by adjusting the firing temperature and time. The thermal properties of WSA-added bricks may be superior to those of standard bricks, as porosity increases and thermal conductivity decreases. This can enhance the thermal insulation performance when used in building products, such as reducing energy consumption for heating and cooling.

4.3.3 Environmental and economic benefits

The use of WSA in brick-making represents a one-dimensional advancement for the environment, as it enables waste valorisation and resource conservation. Life cycle analysis has demonstrated that WSA-containing bricks can reduce their impact by 15–25% compared to conventional fired clay bricks, owing to the extraction of raw materials from the earth, resulting in less extraction of natural resources and lower firing temperatures. Economically, savings are offered on raw materials as well as potential energy during the manufacturing and disposal of WSA. However, economic feasibility relies on the availability of WSA, transportation costs, and market demand for the final products. Quality control and standardisation of the product are necessary to ensure consistency in product quality and market acceptance.

An illustration of the inherent links between water and waste management at the system level is presented in **Figure 3**, with three critical loops identified internally to hydro-economic systems: i) water, ii) energy, and iii) material. This illustrates the

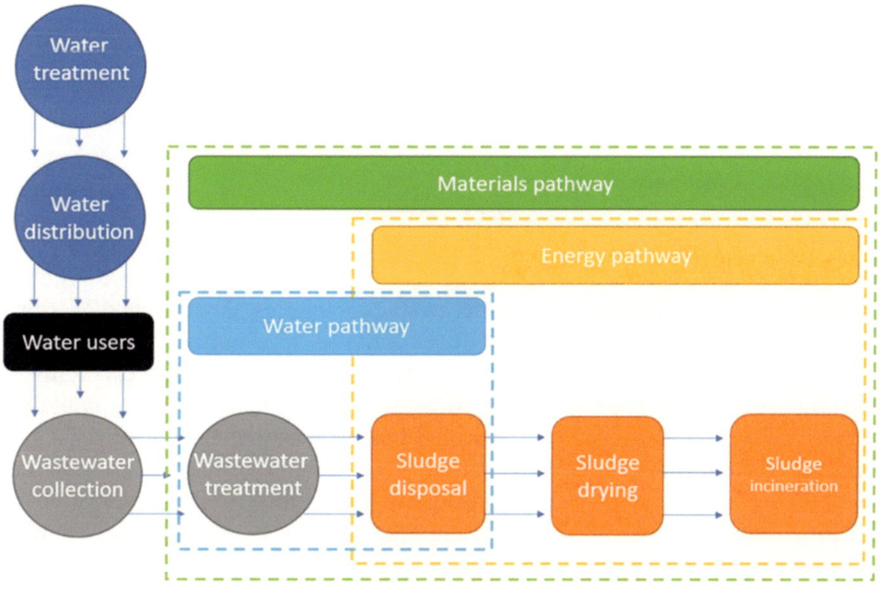

Figure 3.
Combined pathways of wastewater and sludge management systems [52].

water supply chain, from treatment and distribution to the customer, and then the collection and disposal of waste. The sludge generated in these activities is subsequently treated through elimination, drying, and incineration, and then inserted into the material and energy cycles of the system. This indicates that sludge treatment is a key component of waste minimisation and resource/energy harvesting.

4.4 Application in construction materials

In addition to its use as a supplementary cementitious material, WSA can be considered for use in other construction materials, such as asphalt, bricks, blocks, and composites. Different applications with diverse performance requirements and sustainability goals can help unlock the commercial potential of sludge-derived materials. The utilisation of WSA in asphalt mixtures has several advantages, including improved pavement performance and a lower environmental burden. Studies have shown that WSA may substitute for mineral fillers in a traditional mix by 5–15% without significant performance reduction [49]. The filler effects of fine WSA particles and the increased stiffness caused by potential chemical interactions within the asphalt mastic are considered the reasons for the improved performance. By reducing raw material consumption and enhancing properties, the addition of WSA to clay bricks can significantly aid in conserving natural resources [53]. Various studies have found that, in general, the mechanical properties of clay bricks with 10–25% WSA (mass percentage) added to their formulation indicate better performance, but with a lower sintering temperature [54]. As a pozzolana, WSA can serve as a self-strengthening raw material for fired-brick production through non-conventional reactions with clay minerals at low-temperature kilns. The prepared bricks exhibited high strength, low water absorption, and improved durability compared to traditional clay bricks. The WMQM concrete masonry units showed the same trend in terms of strength and durability properties. Up to 15–30% replacement of cement in concrete block production can be achieved using WSA, thereby maintaining or improving the block's performance. Advanced composite materials incorporating WSA, including geopolymer composites and lightweight construction materials, offer opportunities for high-value applications that maximise the benefits of sludge valorisation.

4.5 Energy and nutrient recovery

Sludge valorisation offers more than material recovery; it also provides significant opportunities for energy and nutrient recovery, which can further the circular economic principles and resource efficiency goals. Anaerobic digestion of sludge produces biogas containing 60–70% methane, representing a substantial renewable energy resource [55]. Large wastewater treatment plants can achieve energy self-sufficiency through biogas utilisation, with surplus energy available for export to the electrical grid [56]. Combined heat and power (CHP) systems can achieve overall energy conversion efficiencies of 75–85%, maximising the value of the recovered energy while providing both electricity and thermal energy for plant operations [56]. Additionally, new technologies for biogas upgrading can achieve biomethane purity levels identical to those of natural gas used in transportation fuels and other applications, which are more economically viable than on-site electricity generation [57, 58].

5. Methodology

This in-depth review of the literature was conducted using a systematic search of peer-reviewed journal articles, conference contributions, and technical reports. The search was conducted in key scientific databases, including Scopus, the Web of Science, and Google Scholar. The search terms used included 'wastewater sludge', 'sewage sludge', 'sludge management', 'resource recovery', and combinations of these words with 'wastewater sludge ash', 'supplementary cementitious materials', 'concrete', 'asphalt', or 'bricks'. The search was narrowed to published works within the previous 10 years to develop a body of more recent studies; however, older foundational literature was also included for historical reference. We included studies in this review based on their relevance, methodological quality, and importance of their conclusions. Data from the selected articles were consequently combined and organised to form an overall picture of sludge reuse.

6. Environmental and regulatory framework

6.1 Environmental impact assessment

The potential environmental impacts of sludge valorisation are systemic, as they may occur during the application or final disposal phases related to treatment processes, toxic emissions and long-term health effects. A comprehensive cradle-to-factory insight is needed to avoid unforeseen trade-offs that might undermine the sustainability goals of valorisation strategies. Among the three options compared in this section, only LCA is suitable for assessing the environmental burden of sludge valorisation systems. The environmental impact of sludge valorisation varies depending on the amount used, treatment characteristics selected, and destination, as well as their mutual relationships and long-term effects. To achieve a net positive impact and avoid negating previous efforts, valorisation strategies should include environmental impact assessments [24, 59].

Among the available assessment tools, LCA is the most suitable framework for evaluating the environmental impacts associated with sludge valorisation systems. The LCA approach considers all stages, from sludge generation through treatment, material production, application, and end-of-life management, enabling the identification of environmental hotspots and optimisation opportunities [96]. The system boundaries for sludge valorisation LCA typically include sludge collection and transport, treatment processes, material production, distribution and application, use-phase impacts, and end-of-life management.

6.2 Heavy metal leaching and safety considerations

Heavy metal contamination is a significant environmental and safety concern associated with sludge valorisation, necessitating comprehensive assessment and management strategies to ensure the safe utilisation of sludge-derived materials. The behaviour of heavy metals during treatment and their long-term stability in construction applications are key factors affecting the feasibility and safety of valorisation approaches. Heavy metals in sludge originate from various sources, including industrial discharges, urban runoff, corrosion of distribution systems, and the use of household products. The common heavy metals of interest are Cu, Zn, Cr, Ni, Pb, Cd, and Hg, which show a wide variation in concentration depending on regional industrial activities and waste minimisation programs currently in operation [60]. The concentrations of heavy metals in municipal sludge have decreased significantly over recent decades due to improvements in industrial pre-treatment and source reduction efforts.

The characterisation of heavy metal speciation in sludge is important because it can lead to different consequences for the environment. For exchangeable and water-soluble metals, binding to organic matter or incorporation into mineral phases reduces their mobility. Treatment processes can alter metal speciation, generally reducing mobility through thermal treatment or stabilisation processes [33]. Thermal treatment of sludges typically reduces the mobility of heavy metals by incorporating them into stable mineral phases and eliminating organic-binding sites.

The assessment of heavy metal leaching from sludge-derived materials requires standardised testing protocols that simulate realistic exposure conditions. The TCLP and European EN 12457 leaching tests are commonly used to evaluate the short-term leaching behaviour under acidic conditions [61]. The long-term leaching behaviour should be evaluated using more advanced test methods, considering weathering, carbonation, and chemical evolution over several years.

Comprehensive risk assessment of heavy metal contaminants in sludge-based materials also needs to consider different exposure routes, namely direct contact, inhalation of dust, and water inundation. One such group of individuals frequently exposed to hazardous substances is construction workers who handle and apply these materials as part of their work. The use of appropriate personal protective equipment and procedures can reduce the risk of exposure, thereby enabling the safe use of sludge-based materials [62].

7. Integration into the circular economy

7.1 Economic viability and market development

The adoption of sludge valorisation processes will rely heavily on economic parameters, not only in terms of direct costs but also considering indirect costs at a multi-stakeholder level. Typical factors that determine the financial viability of a project include the size and scale of the facility in question, the choice of technology employed (incineration or anaerobic digestion), prevailing local market conditions, and renewable energy prices, which factor in price guarantees for power at rates above the market. These depend significantly on the resilience and stability of policy frameworks, as well as locally available incentives for a circular economy. The economic viability of sludge valorisation is multifaceted across various scales, including facility and technology levels, local market conditions, and regulatory landscapes. Thus, it cannot be fully examined using a single method. For large-scale facilities (processing more than 10,000 m^3/yr of sludge), positive overall economic benefits are generally achievable by selling energy and materials, in addition to avoiding disposal costs. Revenue opportunities include biogas sales (at $0.08–0.15 per kWh), sludge ash sales ($20–50 per tonne as a supplementary cementitious material), and tipping fees ($50–150 + per tonne) for organic waste co-processing [63, 64]. However, product marketing is challenging due to the variability in product quality, unstable supply chains, and issues with customer acceptance. Increasing environmental consciousness, regulatory constraints, and the transition to a circular economy are believed to offer opportunities for both pioneer and policy-driven sludge-derived material markets [65, 66]. Similarly, implementing quality assurance protocols, standardised specifications, and certification systems can help mitigate the market risks faced by land developers when introducing new users.

7.2 Policy and compliance frameworks

Policy instruments are pivotal for the systematic handling of sludge sustainably within the legal framework. These frameworks are designed to help interpreters understand new sludge conditioning and resource-recovery technologies and systems. Increased information and transparency in environmental management are of particular importance to the public; they are likely to oppose the sludge system, which could lead to waste [67, 68]. There is a complex regulatory environment surrounding the valorisation of sludge; however, this can only be achieved through understanding regulations and engaging with regulatory bodies throughout the process to ensure compliance and health safety. The policy landscape is favouring the circular economy and resource recovery, with regulatory trends encouraging valorisation rather than traditional disposal options.

Sludge valorisation can also be supported through incentives that link projects to carbon taxation mechanisms or similar external drivers (e.g., extended producer responsibility and circular economic action plans). However, barriers to market deployment, including material compatibility issues, permit acquisition challenges, and a relatively uncertain quality landscape, still exist. Standardisation and the introduction of risk-based regulations could offer significant benefits for broader technology deployment through valorisation technologies.

8. Current challenges and future research directions

8.1 Thermal treatment and ash production

Thermal treatment of sludge provides the most complete volume reduction while enabling energy recovery and ash production for use in construction applications. Incineration, gasification, and pyrolysis are the primary thermal treatment methods, each with its own benefits, and the ash produced varies according to the characteristics of the feedstock. The destruction of organic material and maximum volume reduction (85–95%) are achieved through incineration at 850–900 °C, which also offers potential pozzolana-forming properties. Although the process requires advanced air pollution control systems to comply with environmental standards, the plant can be self-sufficient in terms of energy, producing both steam and electricity. Modern incineration facilities incorporate advanced emission control technologies, such as selective non-catalytic reduction, activated carbon injection, and fabric filtration, to minimise environmental impact.

Gasification processes utilises a limited oxygen supply to partially oxidise organic matter, producing a combustible gas mixture primarily comprising carbon monoxide and hydrogen. Sludge gasification can achieve a higher energy recovery efficiency than direct combustion, while generating ash with different characteristics. However, this stage of the process involves highly sophisticated guidance and gas cleaning systems. The incineration conditions, primarily temperature, residence time, and atmospheric composition, influence the fundamental properties of WSA. Generally, when the incineration temperature is high, the pozzolanic properties of the ashes are more developed because of the formation of additional mineral phases and a lower carbon content. Quality control systems for ash production must monitor both the chemical composition and physical properties to ensure consistent material performance.

Figure 4 shows the life cycle stages underpinning the normalisation and impact of the EPD scheme for construction products from sludge-derived materials. The processes linked to the framework cover both the product level (A1–A3), which involves raw sludge incineration, transport, and ash treatment, and the construction stage (A4–A5), where sludge ash is applied as an additive in concrete or asphalt. Stage (D) considers the utility of products beyond the system boundary, focusing on the reuse, recovery, and recycling of products from ash, employing a circular economy approach and linking it to the overall goal of reducing environmental impact.

8.2 Technical and implementation challenges

Although sludge valorisation technologies have progressively advanced over time, with many technologies moving from laboratory to commercial scales, several technical and implementation bottlenecks still restrict their wider application. Extracting the fullest potential from this low-energy, bulk, and multitasking feedstock – often without losing its identity over multiple recycling cycles – continues to require optimised processing, especially in the effort to achieve consistent material quality at a cost that allows for affordable products. The kinematic viscosity of activated sludge in aeration tanks across different treatment plants varies considerably, influenced by differences in plant design, operational conditions, and seasonal factors, making operational control challenging due to such variability. Therefore, our proposed solution must be robust enough to deliver consistent outputs, regardless of these unpredictable inputs. One of the primary challenges limiting the market application of sludge-derived materials is the need for quality control and standardisation. Improved pre-treatment technologies for sludge facilitate the homogenisation of sludge characteristics before the valorisation step, which should enhance process reliability and product consistency. Market development barriers likely include customer acceptance, supply assurance, and price parity with other competing materials. Robust life-cycle cost analyses and demonstrations or deployment collections that build confidence among end users and investors are required. We will not establish certification or standards for sludge-derived materials in isolation, as this will require collaboration among researchers, industry, and regulatory agencies.

8.3 Future research priorities

Advancing sludge valorisation requires ongoing progress in solving key scientific questions and overcoming the practical barriers to conversion. Some innovative treatment technologies, such as hydrothermal carbonisation, plasma gasification, and supercritical water oxidation, may also provide efficient solutions for further reducing sludge before its use in land application. Research in materials science

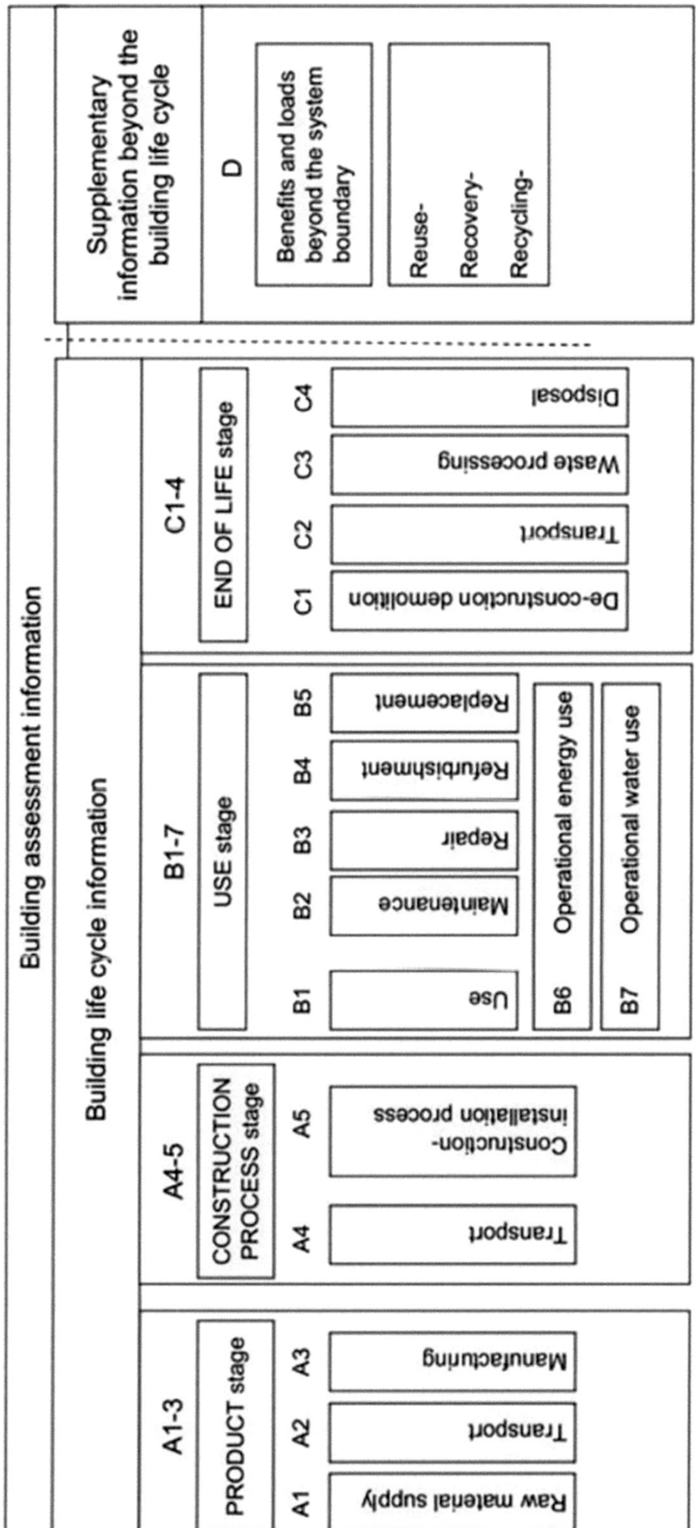

Figure 4.
Life cycle stages of sludge-based material utilisation from sludge generation to material end-use [33, 69].

and engineering has identified high-value uses for sludge-derived components, thereby expanding market opportunities. It is essential to consider environmental safety, risk assessment, and long-term behaviour, including the kinetics of heavy metal leaching, the fate and transport of emerging contaminants, and ecosystem impacts under various environmental conditions. When more widely adopted, digital technologies and integrated process control could enable autonomous valorisation systems to achieve performance optimised for specific products. Research in development policy and economics should identify areas of uncertainty and risk, while also indicating where interventions or market strategies are most effective. International R&D collaborations can accelerate progress. Additionally, sharing high-cost, high-risk R&D capital supports technology transfer and market development across regions.

9. Conclusion and future perspectives

The novel application of sludge, including its conversion into WSA for construction materials, is considered a promising route towards sustainability and a circular economy. From this review, significant progress can be observed in the development of this sustainable material, with numerous studies demonstrating the technical feasibility of utilising WSA as an SCM in concrete, as a filler in asphalt, and as a raw material for brick-making. By adopting WSA in such applications, the need for virgin materials can be reduced, thereby contributing to the preservation of natural resources and a decrease in the carbon footprint of the construction industry.

However, numerous issues must be addressed before WSA can be widely used in construction. One of the main concerns is the variability of WSA's physicochemical properties, which may affect the performance of construction materials in which it is used. A more consistent WSA production is required to ensure uniformity and high-quality products. Further research is necessary to explore the potential of WSA as a beneficial component in various construction applications and to discover innovative ways to utilise this material.

Another significant challenge is the potential release of heavy metals from WSA-bearing construction material. Heavy metal leaching generally remains low, but this issue must be managed carefully. It is recommended that more long-term studies be conducted to assess the environmental implications of WSA utilisation in construction and to identify effective strategies for risk reduction.

From an economic perspective, the application of WSA in construction work is an attractive approach for sludge disposal. However, the financial feasibility of this method is influenced by several factors, including the price of sludge incineration, the market value of WSA, and regulations regarding the incorporation of waste materials. Furthermore, more comprehensive economic evaluations of the cost-effectiveness of WSA in various construction applications are required to determine whether any economic barriers exist to its use.

In summary, using sludge as a raw material for construction could benefit both the environment and the economy. However, several obstacles must be overcome to unlock its potential fully. Further research is needed to standardise the production of WSA, optimise its use in various construction applications, and assess the long-term environmental and economic benefits of this resource for future use. By addressing these issues, there is potential to enhance circularity and sustainability in our economy, transforming sludge from waste into a valuable resource.

Acknowledgments

I would like to thank Tshwane University of Technology for providing a platform to conduct this research.

Conflict of Interest

The authors declare no competing financial interests or personal relationships that could influence the work reported in this study.

Data availability statement

The data supporting these findings are included in this document.

Funding statement

Financial support for this study was received from Tshwane University of Technology, Pretoria, South Africa.

Author details

Kobe Samuel Mojapelo[1*], Williams Kehinde Kupolati[1], Everardt Andre Burger[1], Julius Musyoka Ndambuki[1], Jacques Snyman[1], Emmanuel Rotimi Sadiku[2] and Idowu David Ibrahim[3]

1 Department of Civil Engineering, Tshwane University of Technology, Pretoria, South Africa

2 Institute for NanoEngineering Research (INER) and Department of Chemical, Metallurgy, and Materials Engineering, Tshwane University of Technology, Pretoria, South Africa

3 Department of Industrial Engineering & Operations Management and Mechanical Engineering, Vaal University of Technology, Vanderbijlpark, South Africa

*Address all correspondence to: MojapeloKS@tut.ac.za; sam@smlprojects.co.za

References

[1] Dulaimi A, et al. Application of paper sludge ash and incinerated sewage ash in emulsified asphalt cold mixtures. Frontiers in Materials. 2023;**9**:1074738

[2] Zaharioiu A, et al. Sustainable fuels obtained from the pyrolysis process of sewage sludge. Current Trends in Natural Sciences. 2021;**10**:16–24

[3] Duan B, Feng Q. Risk assessment and potential analysis of the agricultural use of sewage sludge in central shanxi province. International Journal of Environmental Research and Public Health. 2022;**19**(7):4236

[4] Çevik BK. Co combustion of sewage sludge and turkish lignite: The effect of blending on combustion characteristics. Journal of Innovative Engineering and Natural Science. 2023;**1**(1):1–12

[5] Shaddel S, et al. Sustainable sewage sludge management: From current practices to emerging nutrient recovery technologies. Sustainability. 2019;**11**:3435

[6] Wahaab RA, Mahmoud M, Lier JBV. Toward achieving sustainable management of municipal wastewater sludge in Egypt: The current status and future prospective. Renewable and Sustainable Energy Reviews. 2020;**127**:109880

[7] Alevizos V, Georgousis I, Kapodistria A-M. Toward climate neutrality: A comprehensive overview of sustainable operations management, optimization, and wastewater treatment methods. Pollutants. 2023;**3**(4):521–543

[8] Omohwovo EJ. Wastewater management in Africa: Challenges and recommendations. Environmental Health Insights. 2024;**18**:11786302241289681

[9] Ye M, et al. Iron recycle-driven organic capture and sidestream anaerobic membrane bioreactor for revolutionizing bioenergy generation in municipal wastewater treatment. Environmental Science & Technology. 2024;**58**:9350–9360

[10] Buss W. Pyrolysis solves the issue of organic contaminants in sewage sludge while retaining carbon—making the case for sewage sludge treatment via pyrolysis. Acs Sustainable Chemistry & Engineering. 2021;**9**(30):10048–10053

[11] Hale SE, et al. the reuse of excavated soils from construction and demolition projects: Limitations and possibilities. Sustainability. 2021;**13**:6083

[12] Chamath DYYAD, Babel S. Feasibility of a self-sustaining microbial fuel cell incorporating natural processes of bacteria and microalgae towards a circular bioeconomy. Applied Research. 2023;**3**(2):e202200127

[13] Syafrudin S, et al. Feasibility study for mining waste materials as sustainable compost raw material toward enhanced landfill mining. Polish Journal of Environmental Studies. 2023;**32**:2323–2335

[14] Azzahra F, et al. Feasibility study of processing wood twig waste into liquid smoke at the integrated waste management facility Diponegoro University. IOP Conference Series: Earth and Environmental Science. 2024;**1414**:012087

[15] Nurgaliev T, Koshelev VM, Müller J. Simulation model for biogas project efficiency maximization. Bioenergy Research. 2022;**16**(2):1084–1098

[16] Montero-Guadarrama I, et al. Reverse osmosis coupled with ozonation for clean water recovery from an industrial effluent: Technical and economic analyses. Membranes. 2025;**15**:33

[17] Đurđević D, Žiković S, Blecich P. Sustainable sewage sludge management technologies selection based on techno-economic-environmental criteria: Case study of croatia. Energies. 2022;**15**(11):3941

[18] Brunner PH, Morf LS. Waste to energy, indispensable cornerstone for circular economy: A mini-review. Waste Management & Research: The Journal for a Sustainable Circular Economy. 2024;**43**(1):26–38

[19] Cecconet D, Capodaglio AG. Sewage sludge biorefinery for circular economy. Sustainability. 2022;**14**(22):14841

[20] Kwapiński W, Kolinovic I, Leahy JJ. Sewage sludge thermal treatment technologies with a focus on phosphorus recovery: A review. Waste and Biomass Valorization. 2021;**12**(11):5837–5852

[21] Feng J, et al. Assessing metal contamination and speciation in sewage sludge: Implications for soil application and environmental risk. Reviews in Environmental Science and Bio/Technology. 2023;**22**:1037–1058

[22] Rathika K, Kumar S, Yadav BR. From pollutant to powerhouse: The untapped potential of sewage sludge and wastewater. EMBO Reports. 2023;**24**(12):e58201

[23] Haranghy L, et al. Detection of efficiency of microwave-enhanced sludge treatments by dielectric measurements. Analecta Technica Szegedinensia. 2021;**15**:53–57

[24] Patel K, Singh SK. A life cycle approach to environmental assessment of wastewater and sludge treatment processes. Water and Environment Journal. 2022;**36**(3):412–424

[25] Apollo S. A review of sludge production in South Africa municipal wastewater treatment plants, analysis of handling cost and potential minimization methods. Physical Sciences Reviews. 2023;**8**(11):3847–3857

[26] Mojapelo K, et al. Durability and environmental impact of wastewater sludge ash as a cement replacement in concrete: Challenges and future directions. Materials Circular Economy. 2025;7:1–19

[27] Rutkowska G, et al. Modification of ordinary concrete using fly ash from combustion of municipal sewage sludge. Materials. 2020;**13**:487

[28] Tantawy M, et al. Evaluation of the pozzolanic activity of sewage sludge ash. International Scholarly Research Notices. 2012;**2012**:487037

[29] He P, et al. The mechanism of supplementary cementitious materials enhancing the water resistance of magnesium oxychloride cement (MOC): A comparison between pulverized fuel ash and incinerated sewage sludge ash. Cement and Concrete Composites. 2020;**109**:103562

[30] García-Díaz A, et al. Effect of the incorporation of spent diatomaceous earths on the properties of alkaline activation cements based on sewage

sludge ash. Archives of Civil and Mechanical Engineering. 2024;**24**:62

[31] Hassooni S, Ethaib S. Evaluation the effect of reuse sewage sludge and sewage sludge ash on concrete for cement replacement. Journal of Engineering Science and Technology. 2020;**15**(5):3404–3418

[32] Sun K, et al. Recycling of incinerated sewage sludge ash and waste glass powder in alkali-activated slag for sewer rehabilitation. Materials and Structures. 2024;**57**:88

[33] Haustein E, et al. Influence of cement replacement with sewage sludge ash (SSA) on the heat of hydration of cement mortar. Materials. 2022;**15**:1547

[34] Latosińska J. Risk assessment of soil contamination with heavy metals from sewage sludge and ash after its incineration. TC. 2020;**1**(1):3

[35] Chen Z, Poon CS. Comparing the use of sewage sludge ash and glass powder in cement mortars. Environmental Technology. 2017;**38** (11):1390–1398

[36] Yu R, et al. Utilizing CaCl2 to promote the enrichment and bioavailability of phosphorus in incinerated sludge ash. Water Science & Technology. 2023;**87**:2634–2647

[37] Wang L, et al. A mini-review of sludge-derived biochar (SDB) for wastewater treatment: Recent advances in 2020–2025. Applied Sciences. 2025;**15**:6173

[38] Basri S, et al. Beyond conventional approaches: Sustainable valorization of sewage sludge-challenges and opportunities. Sustainable Materials and Technologies. 2025;**45**:e01496

[39] Panahi A, et al. A systems-based approach to circular sludge management: Data-driven foresight, sustainability assessment, and strategic evaluation. Journal of Environmental Management. 2025;**392**:126615

[40] García-López CD, et al. Evaluating sludge management strategies using wastewater treatment plant simulations and life cycle assessment. Journal of Material Cycles and Waste Management. 2025;**27**(5):2972–2986

[41] Kacprzak M, Baran J, Fijalkowski K. Emerging sewage sludge treatment technologies for land carbon sequestration: A comprehensive review. Journal of Material Cycles and Waste Management. 2025;**27**(5):2866–2886

[42] Salva J, et al. Analysis of the current state of sewage sludge treatment from the perspective of current European directives. Environmental Sciences Europe. 2025;**37**:1–27

[43] Kim D, Phae CG. Analysis of the effect of sludge reduction and energy production from introducing anaerobic digest plant into a sewage treatment plant. Water. 2022;**14**(11):1821

[44] Giwa AS, et al. Advances in sewage sludge application and treatment: Process integration of plasma pyrolysis and anaerobic digestion with the resource recovery. Heliyon. 2023;**9**(9): e19765

[45] Fini EH, et al. Perspectives on innovative non-fertilizer applications of sewage sludge for mitigating environmental and health hazards. Communications Engineering. 2024;**3**:178

[46] Memon SA, et al. Incorporation of wheat straw ash as partial sand

replacement for production of eco-friendly concrete. Materials. 2021;**14**:2078

[47] Rafiza AR, et al. The physical and mechanical properties of autoclaved aerated concrete (AAC) with recycled AAC as a partial replacement for sand. Buildings. 2022;**12**:60

[48] Stempkowska A, Gawenda T. Artificial lightweight aggregate made from alternative and waste raw materials, hardened using the hybrid method. Scientific Reports. 2024;**14**(1):16880

[49] Mojapelo KS, et al. Performance evaluation and future prospects of the wastewater sludge ash as a filler material in asphalt mixture. Materials Circular Economy. 2025;7(1):16

[50] Azarhoosh A, Asadi AH. Effect of using sewage sludge ash on the fatigue performance of mastic and asphalt mixture. Case Studies in Construction Materials. 2025;**23**:e05115

[51] Nadupuru SR, et al. Utilizing Sewage Sludge Ash as an Eco-Friendly Filler in Bituminous Mixture. In: E3S Web of Conferences. 2025. EDP Sciences

[52] Smol M. Circular economy in wastewater treatment plant—water, energy and raw materials recovery. Energies. 2023;**16**(9):3911

[53] Katman HY, et al. Workability, strength, modulus of elasticity, and permeability feature of wheat straw ash-incorporated hydraulic cement concrete. Buildings. 2022;**12**:1363

[54] Kore SD, et al. Concrete Made From Waste Paper sludge (WPS): A sustainable Material. Technological Sustainability. 2023;**2**:188–205

[55] Calabrò PS, et al. A short bibliographic review concerning biomethane production from wastewater sludge. Waste Management & Research: The Journal for a Sustainable Circular Economy. 2024;**43**:297–305

[56] Capodaglio AG, Callegari A. Energy use and recovery in wastewater treatment facilities. Renewable Energy & Power Quality Journal. 2024;**18**:425–430

[57] Wang B, et al. A novel district heating system coupled with reclaimed water source heat pump: Energy, economic, environmental, and application analysis. Energy Technology. 2022;**10**(10):2100831

[58] Neugebauer G, et al. Potentials and integrated suitability pre-assessment of wastewater treatment plants as local energy cells. Frontiers in Environmental Science. 2022;**9**:785557

[59] Kathi S, et al. Wastewater and sludge valorisation: A novel approach for treatment and resource recovery to achieve circular economy concept. Frontiers in Chemical Engineering. 2023;**5**:129783

[60] Tytła M, Widziewicz K. Heavy metals in municipal sewage sludge – A brief characteristic of potential threats and methods used to assess the ecological risk. Environment, Earth and Ecology. 2021;**5**(1):18–25

[61] Zhang X, et al. Identification of industrial sewage sludge based on heavy metal profiles: A case study of printing and dyeing industry. Environmental Science and Pollution Research. 2021;**29**:12377–12386

[62] Samaila B, Kalgo ZM, Maidamma B. Exposure to heavy metals in oil and gas wastes: A systematic

review on health hazards assessment and mitigation in Nigeria. Advanced Research in Medical and Health Sciences. 2023;**1**(1):22–32

[63] Kowalski Z, et al. Conversion of sewage sludge into biofuels via different pathways and their use in agriculture: A comprehensive review. Energies. 2024;**17**:1383

[64] Tooraj M, et al. Future of Sludge Management. In: Wastewater Treatment and Sludge Management Systems – The Gutter-to-good Approaches. IntechOpen; 2024

[65] Zhang J, et al. Environmental and economic implication of implementation scale of sewage sludge recycling systems considering carbon trading price. Sustainability. 2022;**14**:8684

[66] Morsy O, et al. A superstructure mixed-integer nonlinear programming optimization for the optimal processing pathway selection of sludge-to-energy technologies. Sustainability. 2023;**15**:4023

[67] Ejairu U, et al. Eco-friendly wastewater treatment technologies (Concept): Conceptualizing advanced, sustainable wastewater treatment designs for industrial and municipal applications. Journal of Environmental Sustainability. 2024;**2**:083–104

[68] Kannan R, et al. Smart sensor-based analysis of turbidity variations during waste water treatment: A review. Annual Research & Review in Biology. 2025;**40**:7–20

[69] EN B. Sustainability of construction works assessment of environmental performance of buildings calculation method. BS EN. 2011;**15978**:2011